Uncertainty in Mechanical Engineering

Edited by
Holger Hanselka
Peter Groche
Roland Platz

Uncertainty in Mechanical Engineering

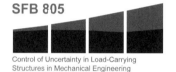

Selected, peer reviewed papers from the
1st International Conference on
Uncertainty in Mechanical Engineering
(ICUME 2011),
November 14-15, 2011, Darmstadt, Germany

Edited by

**Holger Hanselka, Peter Groche
and Roland Platz**

Copyright © 2012 Trans Tech Publications Ltd, Switzerland

All rights reserved. No part of the contents of this publication may be reproduced or transmitted in any form or by any means without the written permission of the publisher.

Trans Tech Publications Ltd
Kreuzstrasse 10
CH-8635 Durnten-Zurich
Switzerland
http://www.ttp.net

Volume 104 of
Applied Mechanics and Materials
ISSN 1660-9336

Full text available online at http://www.scientific.net

Distributed worldwide by	and in the Americas by
Trans Tech Publications Ltd	Trans Tech Publications Inc.
Kreuzstrasse 10	PO Box 699, May Street
CH-8635 Durnten-Zuerich	Enfield, NH 03748
Switzerland	USA
	Phone: +1 (603) 632-7377
Fax: +41 (44) 922 10 33	Fax: +1 (603) 632-5611
e-mail: sales@ttp.net	e-mail: sales-usa@ttp.net

printed in Germany

Preface – Control of Uncertainty in Mechanical Engineering

During its lifetime, any mechanical engineering product will go through various phases in the product development, production and usage. Phases in development extend from the initial idea, product conception and design to the release of the finished product for serial production. These phases can be described as serial and/or parallel virtual process chains. Process chains become physical and real when the product is fabricated and used. In the physical process chain, phases extend from the production of the raw material, manufacture and use of the product right up to its re-use or disposal. Uncertainty occurs in all phases, which has a critical influence on process properties and consequently on product properties. In particular in the case of products with load-carrying function, incorrect assessment due to uncertainty may have catastrophic consequences in terms of safety and profitability of the product, [1] and [2].

The objective of controlling uncertainty in mechanical engineering is to significantly enhancing safety, reliability and economic efficiency in development, production and use, and conserving natural resources. Essentially, uncertainty has to be taken for granted; however, its influence during the product lifetime – from the material/semi-finished product through production and usage right up to its re-use – and during the product lifecycle – from market introduction, through growth and saturation right up to its decline – can be controlled and hence minimized. Particularly in the area of light-weight construction, high demands are made with regard to efficiency – such as low weight and low production cost with adequate load-bearing capacity. This means that in this field, and above all, the control of uncertainty is of particular importance. It is also the focus of the research conducted by the Collaborative Research Centre SFB 805 at Technische Universität Darmstadt, Germany, host of the 1st International Conference on Uncertainty in Mechanical Engineering ICUME 2011 and funded by the Deutsche Forschungsgemeinschaft DFG.

As a first step to reach the goals mentioned above, known methods and technologies for the development, production and utilisation of load-carrying systems, up until their re-use need to be evaluated, with regard to their uncertainty potential. Based on this information, uncertainty can be described in process models and assessed, so that it can eventually be controlled with the help of new methods and technologies to be developed.

1. Advanced methods of robust design, mathematical optimisation methods for robust product design and mathematical models for the combination of active and passive load-carrying components within a system network, appropriate information models for representation and visualisation of uncertainties and new assessment methods will be developed in product development.

2. In production, process chains will be optimised with the help of the mathematical methods described in above. Metal-forming and metal-cutting methods will be rendered more flexible whilst maintaining a consistent level of production quality. Functional materials for active components will be integrated at an early stage.

3. During the use of the load-bearing system, new usage monitoring methods will ensure the permanent acquisition of actual loads, whilst advanced mechatronic and adaptronic or adaptive technologies will stabilise and attenuate the load-carrying structure.

Finally, structure/property relationships derived from the usage process may provide information on the quality and suitability of the product under actual conditions of use – with feedback into development and production.

There are numerous definitions and descriptions of uncertainty, depending on the area of sciences, e.g. data uncertainty, measurement uncertainty, uncertainty of information and many more. However, most definitions do not fully cover combinations and dependencies of uncertainty throughout processes of a product's full life time. Especially when providing high sensitive safety aspects, it is important to know how uncertainty propagates from one process to another, like from processes of development to production, from production to use and, if necessary to reuse. Well known approaches that classify uncertainty are worth to be mentioned:

- Knetsch [3] for example distinguishes between aleatoric and epistemic uncertainty. Aleatoric uncertainty includes random variation of parameters which can not be further reduced in general [4]. Aleatoric uncertainty mainly appears in real product life cycles. Epistemic uncertainty is caused by insufficient information about the product or the process. Due to lack of knowledge, reality is depicted incompletely, incorrectly and insufficiently detailed.

- Uncertainty could be classified into four types, depending on the ability of making a statement about probabilities and the ability to determine a characteristic consequence as well as a meaning of uncertainty [5] and [6]. If probability and consequences are known, the uncertainty is stochastic and therefore well controllable. However, if probability is known but the consequences are not, there is a certain element of ambiguity. If the consequences are known but the probability is not, it is a matter of simple uncertainty. If neither probability nor consequences are known, ignorance prevails.

- International standards like DIN-EN-ISO 12100–1 [7] for machine safety or 2006/42/EG [8] for machine standards define the term uncertainty as
  - a safety function of a machine that, in case of failure, may increase different risks,
  - reliability of a machine to fulfil a certain task for a certain time without failure,
  - risk in combination of probability and the extend of failure as well as
  - safety components, faults, failure etc.

  The term uncertainty is mostly used in an indirect manner in the field of measuring uncertainty or legal uncertainty.

However, a practical, consistent, systematic and comprehensive description and evaluation of uncertainty as well as, eventually, a way to control uncertainty throughout the product's lifetime are still missing. Therefore, the SFB 805 formulated the following working hypothesis describing uncertainty: *Uncertainty occurs when process properties of a system can not, or only partially be determined* [1]. According to this hypothesis, the SFB 805 developed a model of uncertainty to provide a reasonable categorization of uncertainty properties into: unknown uncertainty, estimated uncertainty and stochastic uncertainty. No knowledge or unknown uncertainty occurs when effects and resulting deviation of a regarded property of uncertain processes are unknown. Based on this state of knowledge, no decisions can be made to control uncertainty. With little knowledge or estimated uncertainty, the probability distribution of the resulting deviation is only known partially. Stochastic uncertainty occurs when the effects and the resulting deviations of a considered uncertain property are sufficiently (ideally completely) described by a probability distribution. Stochastic uncertainty is present after extensive analyses of properties in terms of quantifiable experiments and measurements. While differentiating between the three categories, no sharp boundary can be drawn. The transitions between the categories are fluent. As a general rule, unknown uncertainty becomes stochastic uncertainty if the amount of available and secure information increases.

The Organizing committee of the first International Conference on Uncertainty in Mechanical Engineering – ICUME is pleased to present several works from an international community and from the SFB 805 giving an academic and industrial perspective to describe, evaluate and to control uncertainty in:

1 Development
2 Production and
3 Usage.

The editors hope to meet the interest of a broad readership with the selection of the following contributions and like to motivate for further investigations.

Holger Hanselka, Peter Groche and Roland Platz

About ICUME

The aim of ICUME is to discuss methods and technologies to describe, evaluate and control uncertainty in mechanical engineering applications. International scholars and specialists come together to provide a broad forum to discuss the description, evaluation, avoidance, elimination of and adaptation to uncertainty in planning, development, production and usage of mechanical structures, systems and machines throughout their complete lifetime. Engineers, mathematicians and other areas of expertise working in uncertainty evaluation exchange latest research results and application of uncertainty control.

The proceedings show some new approaches and examinations for controlling uncertainty in mechanical engineering. By controlling uncertainty, safety margins between mechanical loading and strength will be lowered, oversizing will be reduced, resources will be preserved, range of application will be widened and economic advantages will be achieved.

The Organizing Committee likes to thank the Deutsche Forschungsgemeinschaft DFG for funding the Collaborative Research Centre SFB 805 at Technische Universität Darmstadt, Germany, and for helping to realize ICUME.

Local Organizing Committee

E. Abele	Production Engineering and Cutting Machine Tools, Technische Universität Darmstadt, Germany
R. Anderl	Computer Integrated Design, Technische Universität Darmstadt, Germany
H. Birkhofer	Product Development and Machine Elements pmd, Technische Universität Darmstadt, Germany
A. Bohn	Product Development and Machine Elements pmd, Technische Universität Darmstadt, Germany
P. Groche	Production Engineering and Forming Machines, Technische Universität Darmstadt, Germany, *Conference Co-Chair*
H. Hanselka	System Reliability and Machine Acoustics, Technische Universität Darmstadt, Germany and Fraunhofer Institute of Structural Durability and System Reliability LBF, Germany, *Conference Chair and Head of SFB 805*
H. Kloberdanz	Product Development and Machine Elements pmd, Technische Universität Darmstadt, Germany
U. Lorenz	Mathematical Optimization, Technische Universität Darmstadt, Germany
P. Pelz	Fluid Systems Technology, Technische Universität Darmstadt, Germany
R. Platz	Fraunhofer Institute of Structural Durability and System Reliability LBF, Germany
S. Ulbrich	Mathematical Optimization, Technische Universität Darmstadt, Germany

International Scientific Committee

J. Allwood	Department of Engineering, Low Carbon Materials Processing, University of Cambridge, UK
G. Diana	Dipatermento di Meccanica, Politecnico di Milano, Italy
P. Göransson	Department of Aeronautics and Vehicle Engineering Royal Institute of Technology, Sweden
E. Macha	Department of Mechanics and Machine Design, Technical University of Opole, Poland
A. Plummer	Department of Mechanical Engineering, Centre for Power Transmission and Motion Control University of Bath, UK

References

[1] Hanselka, H., Platz, R.: Ansätze und Maßnahmen zur Beherrschung von Unsicherheit in lasttragenden Systemen des Maschinenbaus (Controlling Uncertainties in Load Carrying Systems), VDI-Zeitschrift Konstruktion, Ausgabe November/Dezember 11/12-2010, (2010), S. 55-62

[2] Engelhardt, R., Enss, G., Koenen, J., Sichau, A., Platz, R., Kloberdanz, H., Birkhofer, H., Hanselka, H.: A Model to Categorise Uncertainty in Load-Carrying Systems, Conference MMEP Modelling and Management of Engineering Processes, 19. - 20.07.2010 in Cambridge/UK, (2010), pp. 53-64

[3] Knetsch, T.: Unsicherheiten in Ingenieurberechnungen (Uncertainty in Engineering Calculation), Shaker-Verlag, Aachen, 2004.

[4] Chalupnik, M., J., Wynn, D., C., Clarkson, P., J.: Approaches to Mitigate the impact of uncertainty in development processes, In: Proceedings of the 17th International Conference on Engineering Design, Stanford, 2009, pp.1-459 -470.

[5] Andrews, C.J., Hassenzahl, D.M., Johnson, B.B.: Accommodating Uncertainty in Comparative Risk, Risk Analysis, Vol. 24, No. 5, 2004.

[6] Stirling, A.: Risk, uncertainty and precoution: Some instrumental implications from the social sciences, Negotiation Change, 2003.

[7] DIN EN ISO 12100–1, Safety of machinery - Basic concepts, general principles for design - Part 1: Basic terminology, methodology - Amendment 1 (ISO 12100-1:2003/Amd 1:2009)

[8] 2006/42/EG – Directive of the European Parliament and Council from 17.05.2006 for Machines, change of Directive 95/16/EG (revised version). In: Official Register of the European Union, 09.06.2006

Table of Contents

Preface v
Committees ix
References x

1. Uncertainty in Development

Behaviour Prediction Framework in System Architecture Development
K. Osman, M. Štorga, T. Stanković and D. Marjanović 3

A Second Order Approximation Technique for Robust Shape Optimization
A. Sichau and S. Ulbrich 13

Approaches for Assessment of Non-Determinism in Different Stages of Engineering Product Design Using Finite Element Analysis
D. Vandepitte and D. Moens 23

An Approach to Classify Methods to Control Uncertainty in Load-Carrying Structures
R. Engelhardt, J.F. Koenen, M. Brenneis, H. Kloberdanz and A. Bohn 33

Time-Dependent Fuzzy Stochastic Reliability Analysis of Structures
W. Graf and J.U. Sickert 45

Ontology-Based Information Model for the Exchange of Uncertainty in Load Carrying Structures
A. Sprenger, M. Haydn, S. Ondoua, L. Mosch and R. Anderl 55

An Approach of Design Methodology and Tolerance Optimization in the Early Development Stage to Achieve Robust Systems
P. Steinle and M. Bohn 67

2. Uncertainty in Production

Influence of Tolerances on Mechatronic Comfort Systems Behavior
M. Bohn and F. Wuttke 75

Control of Uncertainties in Metal Forming by Applications of Higher Flexibility Dimensions
P. Groche, M. Kraft, S. Schmitt, S. Calmano, U. Lorenz and T. Ederer 83

Early Stage Geometrical Deviation Optimization – An Automotive Example for Sheet Metal Parts
M. Bohn, P. Steinle and F. Wuttke 95

Methods for the Control of Uncertainty in Multilevel Process Chains Using the Example of Drilling/Reaming
M. Haydn, T. Hauer and E. Abele 103

Integration of Smart Materials by Incremental Forming
M. Brenneis, M. Türk and P. Groche 115

3. Uncertainty in Usage

Fatigue Life Estimation under Cyclic Loading Including Out-of-Parallelism of the Characteristics
M. Kurek and T. Łagoda ... 125

Approach for a Consistent Description of Uncertainty in Process Chains of Load Carrying Mechanical Systems
T. Eifler, G.C. Enss, M. Haydn, L. Mosch, R. Platz and H. Hanselka ... 133

Assessment of Uncertainty for Structural and Mechatronics Engineering Applications
S. Donders, L. Farkas, M. Hack, H. Van der Auweraer, R. d'Ippolito, D. Moens and W. Desmet ... 145

Uncertainties with Respect to Active Vibration Control
P.F. Pelz, T. Bedarff and J. Mathias ... 161

Effect of Suspension Parameter Uncertainty on the Dynamic Behaviour of Railway Vehicles
L. Mazzola and S. Bruni ... 177

Evaluation and Control of Uncertainty in Using an Active Column System
J.F. Koenen, G.C. Enss, S. Ondoua, R. Platz and H. Hanselka ... 187

Influence of the Selected Fatigue Characteristics of the Material on Calculated Fatigue Life under Variable Amplitude Loading
A. Niesłony and A. Kurek ... 197

Keyword Index ... 207

Author Index ... 209

1. Uncertainty in Development

Behaviour Prediction Framework in System Architecture Development

Krešimir Osman[1,a], Mario Štorga[1,b], Tino Stanković[1,c], Dorian Marjanović[1,d]

[1]Faculty of Mechanical Engineering and Naval Architecture, University of Zagreb, Ivana Lučića 5, 10000 Zagreb, Croatia

[a]kresimir.osman@fsb.hr, [b]mario.storga@fsb.hr, [c]tino.stankovic@fsb.hr, [d]dorian.marjanovic@fsb.hr

Keywords: behaviour prediction framework, structural complexity management, model predictive control, managing uncertainty

Abstract This paper proposes a Behaviour Prediction Framework with an objective to help designers tackling the problem of uncertainty emerging from system architecture and the effects of the uncertain operating conditions. The proposed framework combines structural and dynamic system model. The Design Structure Matrix is applied to model structural arrangements and dependencies between the subsystems. The Model Predictive Control is applied to model the system in discrete and continuous dynamic domains. As the result of the proposed framework, stability analysis of subsystems in interaction become possible and feedback on system architecture could be provided. To test validity of the proposed approach, the test case involving climate chamber with heat regeneration is presented.

Introduction

Design as an activity is based on the principle of generation and testing solution alternatives until they conform to designer's understanding of what has to be designed. At any abstraction level used during the design process two aspects interfere: establishment of the system architecture and evaluation the system behaviour which emerged as the result of the proposed architectural structure [1, 2]. Thus, the system's performance is dependent on designer's understanding of a design problem including personal beliefs and experience, and on the emergent behaviour of the system which was designed to perform within certain acceptable limits. Achieved by a designer, behaviour of a technical system may reflect only aspects and traits of behaviour modes which can emerge from established system's architecture introducing in such way uncertainty in the respect to the system functionality. In the literature the uncertainty is defined as a state of having limited knowledge where it is impossible to exactly describe existing state or future outcome, more than one possible outcome [3]. From complex system's research area [4] it is known that even a small change into system's architecture can lead to unexpected or even unstable behaviour of the whole system, or likewise that small perturbation of input conditions as unforeseen environmental conditions or modes of use yield in an undesired system's behaviour. In simple cases uncertainty arising from system environment can be handled by estimating the probabilities of such events or can be handled by use feedback to correct for unexpected or incorrectly predicted environment changes.

This work presents Behaviour Prediction Framework which is aimed to provide designer with a feedback about expected behaviour of predefined subsystem architecture in two cases: under the given operating conditions and in a case of their unexpected change. Provided feedback should point out the elements within system architecture which are not able to operate within given parameters thus causing unstable system behaviour. Although the long term research goal is to establish automated feedback between structural and behavioural domains, at the current research stage the transformation between domains is performed manually, as well as resolving the implications of simulation results to system's architecture.

The background of this research in the following section will present related work which attempt at unify structural and dynamic system models. Third section will provide more detailed description of the behaviour prediction methodology. Section four will provide description of Behaviour Prediction Framework what is followed by a case study with a goal to evaluate applicability of the proposed framework. Discussion on obtained results and conclusions close this paper.

Combining Structural and Dynamic System Models

There are two meaningful kinds of complexity relevant for the modern technical systems: structural and behavioural complexity. Modelling the system's structure and its dynamics within the same methodology is gaining importance as it improves the system's understanding. Two domains are usually considered as separate issues: structural complexity management models cannot describe the system's behaviour and dynamical system modelling methods cannot be applied to large scale or complex systems as it requires detailed and expensive process to acquire information about all interactions. One of the recent approaches to combine two domains is presented in work of Diepold et al. [6] as a Multi-Dynamic Mapping framework (see Figure 1) between Multiple-Domain Matrices [5] and Generalized Hybrid State Model [7].

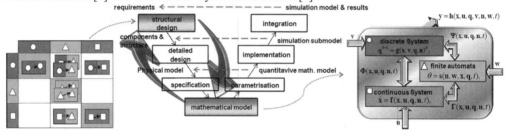

Figure 1. Multi-Dynamic Mapping approach [6]

Multiple-Domain Matrices (MDM) [5] enables different structural views of the system in one model, e.g. requirements, functions, subsystems, components. Method could be used to model the classification of implied domains and dependency types with a goal help designers to keep track of the relevant system aspects and interrelationships. Once all domains are compiled, designers can collect system elements within the domains separately. Such approach allows decomposing, structuring and analysing of the complex systems structural domain.

General Hybrid State Model (GHSM) consists of three dynamical domains [7]: the *discrete subsystem*, which allows the description of discrete-time variable characteristics between system items, the *continuous subsystem* allowing continuous-time variable characteristics and the *finite automation* [8]. The *discrete subsystem* is thereby given by a classical discrete-time dynamical expression and analogous the *continuous subsystem* by ordinary differential equations. The *finite automation* allows a system's theoretical description of system's parts requiring a minimum of knowledge e.g. using Markov chains or Petri nets.

The major interfaces between structural and dynamical system models are constitutionally significant parts of the systems – subsystems that realise key functions. After determining those parts by functional and structural analysis, their behaviour is transformed into a dynamical system representation. The interpretation process transforms the results of the simulation and the findings from the behavioural simulations to answer on the initial problem to be solved. One criterion which is particularly important for the dynamics is the cycle criterion as it is fundamental for feedback loops. Multi-Dynamic Mapping approach use cycles to derive the subsystems, which are significant for the overall system behaviour. However, as there are numerous structural criteria which can be relevant for the system's dynamics, choosing the right criterion for the specific purpose is a critical decision during the structural analysis.

In example given by Diepold et al. [6], the mapping framework is implemented on very simple mechanical system (the ball-pen) in order to illustrate the potential of the proposed models for adjusting the structural design in consequence of the system's performance. In his work [9], Diepold presented structured process modelling approach by extending the DSM to the DynS-DSM (Dynamical System DSM). The properties of structural analysis are thereby kept up during the whole modelling procedure, which results in a discrete-time representation of the system's dynamics. Structural analysis and the effects of structural changes are thus directly transferable into the system's dynamics supporting a coupled system optimization. DynS-DSM approach is extended to a framework (called quad-I/HS), which supports modelling of hybrid dynamical systems [9].

After the analysis of described work we have identified several drawbacks and defined the following possible extensions:

- In contrast to Diepold's mapping framework [6] our approach should offer possibility to mathematically model all of the subsystems behaviour during the operation time. Our assumption is that all subsystems are linear invariant thus represented in state space model, which is suitable for the further system stability analysis.
- System representation in nonlinear form (as in [6]), omits further possibilities to transform the model into linear form. That will be resolved by using Model Predictive Control (MPC) [10, 11] which allows control and adjustment of working parameters both in discrete and continuous domains thus enabling the modelling of the system in real working conditions. Based on the obtained model, stability analysis of subsystems in interaction becomes possible. Thus, relating the MPC methodology to manage uncertainty is a fundamental contribution and advantage of the presented approach.
- Finally, the improvement of the system is conducted directly within system's dynamical behaviour model, thus reducing the design iteration steps.

Model Predictive Control

Model predictive control (MPC) is a methodology originating from the process industry denominating a collection of methods which enable control of constrained linear and non linear systems to meet a desired behaviour [10, 11, 12]. Objective in MPC is understood both as the limit to which the system performance is guided to in order to be economically feasible and as an assurance of the performance stability. The former implies that the limit values have to be known a priori, or to be predicted with a degree of uncertainty if necessary, in order to be able to calculate required signals for corrections. The assurance of the performance stability requires that control must provide precise inputs and effects which will be able to meet on-the-fly uncertainties of the performed system behaviour.

In comparison with conventional control methods which try to rectify the outputs based on the feedback provided as a response to actions undertaken, MPC aims at targeting intended (predicted) behaviour. In order to do so, model a discrete time model of the system is utilized to obtain an estimate of its future behaviour.

Estimation of the future behaviour is accomplished by applying a set of input sequences to a model with measured state/output as initial condition, while taking into account imposed constraints. An optimisation problem built around a performance oriented cost function is then solved to choose an optimal sequence of controls from all feasible sequences as close as possible to desired behaviour. To summarise, the MPC is built based on the following principles [10]:

- The explicit use of a process model for calculating predictions of the system behaviour based on the architecture of considered system.
- The optimisations of an objective function subject to constraints, which yields with control trying to maintain system's stability with optimal performance.
- The receding horizon strategy as a goal behaviour which designer expects from the system.

Directing up to prescribed performance values rather than be directed by past behaviour deviations is observed within system design process when experienced designer produces solutions directed by own in-filed knowledge tin order to meet requirements. Search for set of inputs which are able to maintain system stability by the MPC can tackle the portion of uncertainties showing to the designer that intended behaviour of the system can be kept stable under imposed conditions. Design process congruent features alongside the MPC's the applicability of multivariable problems and ease of use not requiring in-depth control knowledge qualifies MPC based approach as a strong candidate providing robust behavioural system modelling.

Behavior Prediction Framework

The proposed Behaviour Prediction Framework (see Figure 2) is aimed to be used during system architecture design phase for structural and behaviour modelling of the systems. We decided to model system structural complexity in only one domain (subsystems domain) in order to point out the importance of the interaction at this level of system abstraction. By using subsystems domain as a generalisation of a more concrete components domain [13], the functional and geometrical complexity of the system may be reduced to a manageable level. The Design Structure Matrix (DSM) [6, 13, 14] is applied to model arrangements and dependencies between the subsystems. In order to confirm that proposed subsystems arrangements suffice the performance that is expected from the system, the continuous and discrete behaviour domain should be also modelled (see Figure 2) according to general principles of system dynamics.

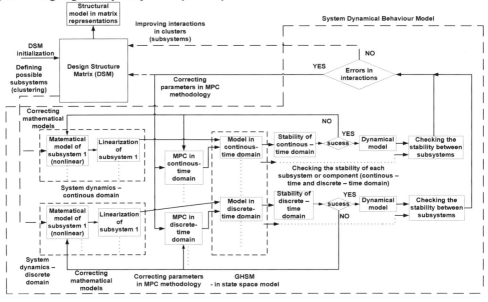

Figure 2. A schema of Behaviour Prediction Framework

The goal of the behaviour modelling is to obtain a linear model in state space representation of the system architecture. The starting point is a non-linear behaviour model of the considered system which may refer to either continuous or discrete time domains. The most common linearization method, which could be applied here, is the expansion in Taylor's series around the equilibrium point [15]. Based on the [16], a high order linear model of dynamical system can be decomposed thus representing it with collections of linear models at different levels of hierarchy. The latter is applied within GHSM [5] module in the framework (Figure 2)

After linearization, the feedback mechanism of the MPC is used to compensate deviations of state variables (system performance) for the predicted equilibrium point. The application of the MPC is to predict the response of the system's output variables relevant for the functionality of the considered system. After applying the MPC the stability is checked for each subsystem for continuous and discrete domain. For energy-based stability analysis Lyapunov method [17] is applied.

After performance stability is checked for each of subsystems, the result may suggest that some of them are unstable under imposed working conditions. In case of instability, a parameters tuning is performed in mathematical model of the subsystem (e.g. air and water volume flow or other parameters which are in correlation), until the subsystem reaches one of the stable condition states (according to [15]). The next step is to check the stability of subsystems that are in interaction. In case of their instability, a refinement of the system architecture within a DSM is required (e.g. we have to add or remove some entities in system architecture and change their relations in order to improve stability of the interactions). The proposed framework will be further explained and illustrated with a case study of climate chamber with heat regeneration that is described in following section.

Case study – climate chamber with heat regeneration

The purpose of the case study is to show how Behaviour Prediction Framework can support designers during conceptual design on the example of the climate chamber. Based on the specification of the initial working conditions the simulations of the chamber subsystems behaviour were performed and accordingly to the feedbacks the final system architecture was proposed.

Climate chamber with heat regeneration is very often an integral part of HVAC for large objects (e.g. shopping malls, hotels or business objects). As within energy management (energy cost) the heat regeneration is very desirable goal, fulfilment of the demand for shorter heating/cooling process time in respect to uncertain environmental conditions is very important. For our particular case study initial working conditions for winter period are given as follows: outdoor (environment) temperature T_o = -10 °C (which is average outdoor temperature for town Zagreb, Croatia) and air flow of q_{va} = 8,5 m³/s. Our goal in the case study was to propose architecture for mentioned working conditions and check the stability of the proposed solution for given working conditions. Also, the proposed architecture should be tested to unexpected working conditions in order to simulate the chamber response on the temperature drop assumed at -30°C, simulating in such way uncertainty of the working conditions that are stochastically happening in Zagreb area.

The first concept of the chamber architecture was developed based on the designer experience with similar systems as presented on Figure 3. This scheme (for initial working conditions) is a starting point for understanding the relationship between main subsystems. A DSM based on Figure 3 is presented on Figure 4.

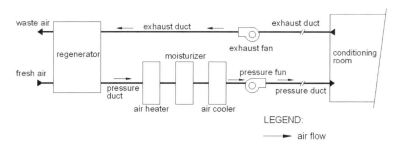

Figure 3. Simplified schema of starting conceptual design of climate chamber

The LOOMEO© (www.teseon.com) was used as a tool for describing climate chamber subsystem structure for further analysis. Possibility for modularization of chamber's subsystems is determined by performing clustering operation over DSM. Figure 4a shows a portion of DSM matrix representation of the architecture after the several steps of refinement including clustering has been conducted. Figure 5b presents DSM in graph representation.

Based on the proposed module clustering [13], the behaviour modelling was conducted as follows. First, the initial proposal of the detailed schema for entire system (climate chamber with heat regeneration) was developed (Figure 5). Three subsystems (two air heaters and moisturizer, please see Figure 6) were selected for performance stability testing both in response to the initially imposed working conditions ($T_o = -10$ °C) and to uncertain working conditions ($T_o = -30$ °C). The selected subsystems were chosen because they initiate the highest change of temperature (energy) in heat chamber (from environment in winter period to conditioning conditions).

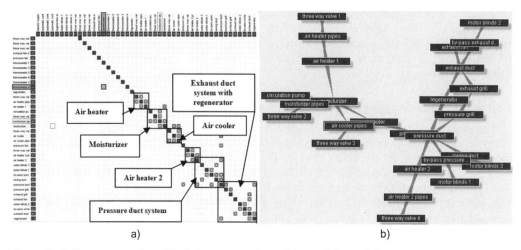

Figure 4. a) Component – based DSM representation with possible modules identified (subsystems), b) Graph representation of system's DSM (screenshots from LOOMEO®)

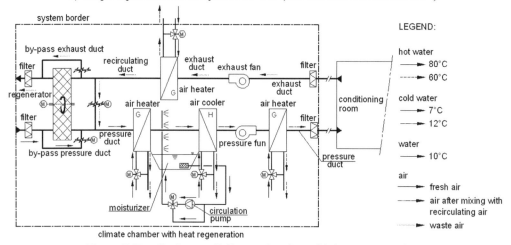

Figure 5. Detail schema of climate chamber with heat regeneration

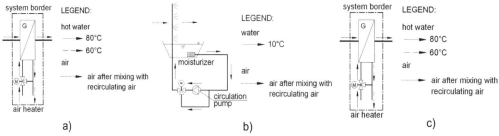

Figure 6. Schema of subsystems: a) subsystem 1 (air heater), b) subsystem 2 (moisturizer), c) subsystem 3 (air heater)

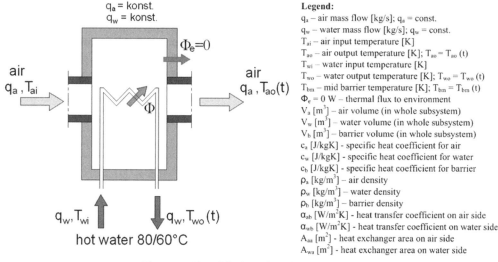

Figure 7. Simplified model of the subsystem 1

The mathematical model (system of differential equations for energy conservation) for subsystem 1 (air heater – Figure 7) in continuous domain are expressed as follows (1-3):

$$\frac{dT_{ao}}{dt} = a_1 \cdot q_{v,a} \cdot (T_{ai} - T_{ao}(t)) - a_2 \cdot (T_{am}(t) - T_{bm}(t)) \tag{1}$$

$$\frac{dT_{wo}}{dt} = c_1 \cdot q_{v,w} \cdot (T_{wi} - T_{wo}(t)) - c_2 \cdot (T_{wm}(t) - T_{bm}(t)) \tag{2}$$

$$\frac{dT_{bm}}{dt} = b_1 \cdot (T_{am}(t) - T_{bm}(t)) - b_2 \cdot (T_{bm}(t) - T_{wm}(t)) \tag{3}$$

where the coefficients $a_1, b_1, c_1, a_2, b_2, c_2$ are given as (4-6):

$$a_1 = \frac{1}{V_a}, \quad a_2 = \frac{\alpha_{ab} \cdot A_{aa}}{\rho_a \cdot c_a \cdot V_a} \tag{4}$$

$$c_1 = \frac{1}{V_w}, \quad c_2 = \frac{\alpha_{wb} \cdot A_{wa}}{\rho_w \cdot c_w \cdot V_w} \tag{5}$$

$$b_1 = \frac{\alpha_{ab} \cdot A_{aa}}{\rho_b \cdot c_b \cdot V_b}, \quad b_2 = \frac{\alpha_{wb} \cdot A_{wa}}{\rho_b \cdot c_b \cdot V_b} \tag{6}$$

The mid air T_{am} and water T_{wm} temperature within air heater are given by the following (7, 8):

$$T_{am}(t) = \frac{T_{ai} + T_{ao}(t)}{2} \tag{7}$$

$$T_{wm}(t) = \frac{T_{wi} + T_{wo}(t)}{2} \tag{8}$$

All parameters from equations (1–8), depend on the properties of fluids (air and water), material (for air heaters), and on air heater design (dimensions of the case plate for air heater and moisturizer). With presented set of equations (1-8) the mathematical model of the subsystems in state space is defined to be used for behaviour simulation in order to check performance stability. For simulation, the subsystem models were described in external scripts (in MATLAB® editor) and imported to the MATLAB Control System Toolbox®, where simulation was performed. The stability was checked with a focus on the internal stability of equilibrium states for homogeneous state equations.

Discussion

Figure 8 illustrates the MPC structure overview in MATLAB® MPC toolbox with two manipulated variables (inputs), three outputs and the results for initial and uncertain working conditions. Manipulated variables in the example are: T_{ai} (input air temperature) and T_{wi} (input water temperature). The outputs from subsystem (*plant*) model are: T_{ao} (output air temperature), T_{wo} (output water temperature) and T_{bm} (middle barrier temperature), which are time dependent variables in our example. Presumed influential factors on the system behaviour for the example in the case study for initial and uncertain conditions are the changes of input temperature to subsystem(s) and air flow though the pressure duct.

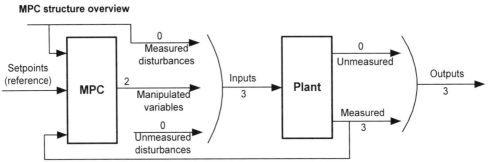

Figure 8. MPC structure overview

The figure 9 depicts simulation outputs for initial working conditions in winter period with input temperature (cca. $T_o = -10$ °C), and increased air flow ($q_{va} = 8,5$ m³/s). On the figure 9a the simulation outputs are shown for improved model after several feedback loops within presented framework were conducted. Improving model assumes change of parameters in mathematical model, adjusting parameters in MPC and changing the system architecture. As said, stability analysis is a part of the proposed framework to check how the considered system will perform under the MPC. The information tells whether the changing and adjusting of parameters was performed in order to reach desired behaviour (i.e. horizon) considering every subsystem and all of the subsystems as a whole. The output architecture is either acceptable or it must be altered to meet the imposed conditions (addition or removal of elements and relations). Figure 9b shows subsystem 1 in asymptotic stability (after the few iteration steps) that confirms a suitable solution for subsystem 1 in respect to initial working conditions.

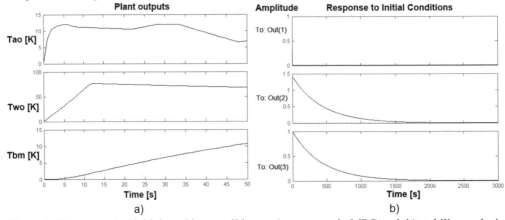

Figure 9. Subsystem 1 at initial working conditions - a) responses in MPC tool, b) stability analysis - asymptotically stable system

Next step is to repeat the simulation and test system in the unexpected working conditions (outdoor temperature drop to -30°C) for system architecture derived on account of the previous simulation. Figure 10 shows simulation outputs for unexpected working conditions with input temperature (cca T_o = -30 °C). On the figure 10a simulation outputs in MPC are shown for model as derived from the previous simulation (model assumes architecture and process parameters as known). Figure 10b present subsystem 1 in an unstable condition. As not being an acceptable solution for subsystem 1, intervention in mathematical model with parameter adjustment is required.

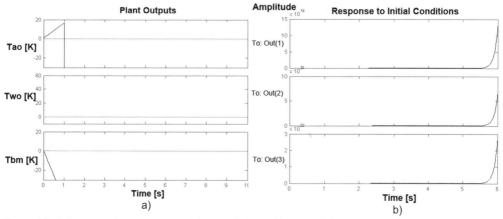

Figure 10. Subsystem 1 at unexpected (uncertain) working conditions - a) responses in MPC tool, b) stability analysis - unstable subsystem

The problem can be resolved by adding these new elements: heat regenerator, recirculation duct and air heater. Every added subsystem assumes a partial role in temperature change thus achieving an increase of the input temperature. Likewise with recirculation duct (and mixing process of fresh air and percent of waste/recirculation air), we decrease air flow throughout the observed system. Starting from the initial system schema (see Figure 3) and refining it throughout proposed framework a climate chamber design concept is obtained which meets both expected and uncertain working conditions.

Conclusion

The paper proposes a Behaviour Prediction Framework which could help designers with uncertainties of system behaviour and system stability prediction during system architecture development stage. The proposed framework offers opportunity to simplify, improve and accelerate development process for systems that are facing uncertain conditions during operating phase. Based on the proposed simulation and prediction method, it is possible to analyse different system architecture arrangements and subsystems interactions against the changes in architecture elements. Framework also enables designers to make refinement on existing subsystem structures, adding new features to them and predicting new behaviour based on the new features.

Future research will be continued in several directions. One of them should be development of the interface between structural and behaviour model, enabling in such way automatic indication of the problems occurring on subsystems level as the result of dynamical analysis. The other possibility is the research on how to apply the extended MPC involving uncertain evolution sets, i.e. the robust MPC [18]. This could allow consideration of an uncertain system under any admissible uncertainty in order to achieve system robust stability.

Acknowledgements

This research is part of funded project "Models and methods of knowledge management in product development" supported by the Ministry of Science and Technology of the Republic of Croatia.

References

[1] V. Hubka, W. E. Eder: *Engineering Design: General Procedural Model of Engineering Design* (Springer – Verlag Berlin Heidelberg, Germany, 1992)

[2] C. T. Hansen, M. M. Andreasen: *Two approaches to synthesis based on the domain theory,* In Engineering Design Synthesis, (ed. A. Chakrabarti), chapter 6, (Springer-Verlag London Limited, UK, 2002), p. 93 – 108.

[3] D. Hubbard: *How to Measure Anything: Finding the Value of Intangibles in Business* (John Wiley & Sons, USA, 2007)

[4] M. Mitchell: *Complexity: A Guided Tour* (Oxford University Press, USA, 2009)

[5] U. Lindemann, M. Maurer and T. Braun: *Structural Complexity Management – An Approach for the Field of Product Design* (Springer – Verlag Berlin Heidelberg, Germany 2009)

[6] K. J. Diepold, W. Biedermann, K. G. M. Eben, S. Kortler, B. Lohmann and U. Lindemann: *Combining Structural Complexity Management and Hybrid Dynamical System Modelling,* In: Proceedings of DESIGN 2010, Dubrovnik, Croatia, volume 2, (2010), p. 1045 - 1054.

[7] M. Buss, M. Glocker, M. Hardt, O. von Stryk, R. Bulirsch, G. Schmidt: *Nonlinear Hybrid Dynamical Systems: Modelling, Optimal Control, and Applications*, In: Analysis and Design of Hybrid Systems - Lecture Notes in Control and Information Science (LNCIS), Springer Berlin, Germany, 2002, p. 311–335.

[8] T.A. Helzinger T.A.: *The Theory of Hybrid Automata* (Berkley, USA, 1996)

[9] K. J. Diepold, F. J. Winkler, B. Lohmann: *Systematical hybrid state modelling of complex dynamical systems: The quad - I/HS framework*, In: Mathematical and Computer Modelling of Dynamical Systems, Vol. 16, No. 4 (2010), p. 347 - 371.

[10] E.F. Camacho and C. Bordons: *Model Predictive Control* (Springer – New York, USA 2004)

[11] L. Wang: *Model Predictive Control System Design and Implementation Using MATLAB,* (Spinger – Verlag London Limited, 2009)

[12] M. Lazar: *Model Predictive Control of Hybrid Systems: Stability and Robustness,* PhD thesis (Technische Universiteit Eindhoven, Eindhoven, 2006)

[13] D. Steward: *The Design Structure Matrix: A Method for Managing the Design of Complex Systems*, In: IEEE Transaction on Engineering Management, Vol. 28, No. 3 (1981), p. 321-342.

[14] T. U. Pimmler, S. D. Eppinger: *Integration Analysis of Product Decompositions.* In: Proceedings of the 1994 ASME-DTM Conference (ASME, Minneapolis, USA, 1994)

[15] R. C. Dorf, R. H. Bishop: *Modern Control Systems* (Pearson Education, Inc., Upper Saddle River, New Jersey, USA, 2011)

[16] W. Stanislawski, M. Rydel: *Hierarchical mathematical models of complex plants on the basis of power boiler example*, In: Archives of Control Sciences, Vol. 20, No. 4, (2010), p. 381–416.

[17] A. Bacciotti, L. Rosier: *Liapunov Functions and Stability in Control Theory*, 2nd edition (Springer - Verlag Berlin Heidelberg, Germany, 2005)

[18] A. Bemporad, M. Morari: *Robust Model Predictive Control: A Survey,* In: Robustness in Identification and Control, Vol. 245, A. Garulli, A. Tesi, A. Vicino (Eds.), Lecture Notes in Control and Information Sciences, Springer -Verlag (1999), pp. 207-226

A Second Order Approximation Technique for Robust Shape Optimization

Adrian Sichau[a] and Stefan Ulbrich[b]

Technische Universität Darmstadt, Department of Mathematics, Nonlinear Optimization,
Dolivostr. 15, D-64293 Darmstadt, Germany

[a]sichau@mathematik.tu-darmstadt.de, [b]ulbrich@mathematik.tu-darmstadt.de

Keywords: robust optimization, applications of robust optimization, robustness, data uncertainty, nonlinear programming, shape optimization, engineering design.

Abstract. We present a second order approximation for the robust counterpart of general uncertain nonlinear programs with state equation given by a partial differential equation. We show how the approximated worst-case functions, which are the essential part of the approximated robust counterpart, can be formulated as trust-region problems that can be solved efficiently using adjoint techniques. Further, we describe how the gradients of the worst-case functions can be computed analytically combining a sensitivity and an adjoint approach. This method is applied to shape optimization in structural mechanics in order to obtain optimal solutions that are robust with respect to uncertainty in acting forces. Numerical results are presented.

1. Introduction

Most real-world applications involve uncertainty in one way or another. Particularly in mechanical engineering the presence of uncertainty is indisputable. In shape optimization of load-carrying structures, for example, one needs to know certain material properties like Young's modulus or Poisson's ratio as well as the forces acting on the considered structure. Because of measurement errors and other technical difficulties the actual values of those parameters usually cannot be determined exactly. Rather, only a nominal value and a maximal deviation are available for the parameters. Similarly it is impossible to implement the design variables exactly as computed, due to e.g. fabrication tolerances.

Since uncertainty may lead to severe economical as well as safety consequences, the subject of controlling the influnece of uncertainty is prevalent in current research. As mathematical optimization plays an important role in design optimization of mechanical structures, it is crucial that the applied methods are capable of incorporating the existence of uncertainty.

Even small perturbations in the parameters of an optimization problem may lead to severe changes in the optimal solution. Furthermore, only a slightly altered optimal solution might already be infeasible for the considered problem. This has been shown in a compelling manner by Ben-Tal and Nemirovski in [1, 2]. Therefore, one has to accept the fact that a nominal solution could turn out to be entirely irrelevant from a practical point of view, when uncertainty in the parameters or in the implementaion of the design variables are present. Consequently, it is reasonable to employ optimization techniques which are capable of providing robust solutions that are immune against the influence of uncertainty.

There are two major approaches to incorporate uncertainty into mathematical optimization problems. Stochastic and Robust Optimization. In Stochastic Programming it is assumed that the regarded uncertainty can be described probabilistically. It is required that the underlying distribution is known or at least can be estimated appropriately. Then, uncertainty can be included using chance constraints, expectation values, or risk measures like for example the value-at-risk. For a detailed picture of the topic we refer to [3, 4, 5] and references therein.

Robust Optimization, on the other hand, assumes that the uncertainty is restricted to a given set. Then one seeks a robust optimal solution that is feasible for all realizations from the uncertainty set and at the same time has the best worst-case performance. The massive interest in robust optimization over the past fifteen years was triggered by the work of Ben-Tal and Nemirovski, e.g. [1, 6, 2] and El Ghaoui et al. [7, 8]. A detailed overview of the current activities can be found in [9, 10] and references therein. It is noticeable that aside the significant advances in robust conic and robust integer programming, there is only isolated work on the robustification of general nonlinear programs, see [11, 12, 13].

In the present paper we adopt and extend the approximation idea proposed in [11, 12]. For the ease of presentation, we describe our approach only for the case of uncertain parameters. We remark, however, that likewise uncertainty in the design variables can be considered and that also uncertainty in the parameters and the design variables can be handled at the same time. Yet, the main idea and the principal techniques remain the same.

The paper is organized as follows. In section 2 we formulate the uncertain optimization problem as well as the uncertainty sets that we want to consider, and also present the associated robust counterpart. In section 3 we describe, how tractable approximations of the robust counterpart can be obtained. First, we sum up the first order approximation of Diehl et al., to subsequently present our second order approach. In section 4 we show that our approximation can be formulated as trust-region problem, and explain how this trust-region problem can be solved. In Section 5 we illustrate how the associated derivatives can be computed. In section 6 we present numerical results for an application example from structural optimization. Section 7 concludes the paper.

2. Problem formulation and uncertainty to be considered

We consider general uncertain nonlinear optimization problems which can be formulated as

$$\min_{y \in Y,\ x \in X} J(y, x, p) \quad \text{s.t.} \quad G(y, x, p) \leq 0,\ C(y, x, p) = 0, \tag{1}$$

where $x \in X$ denotes the design variable and $y \in Y$ is called the state variable. We will focus only on finite-dimensional problems, and therefore set $X = \mathbb{R}^{n_x}$ and $Y = \mathbb{R}^{n_y}$. Furthermore, $p \in \mathbb{R}^{n_p}$ represents an uncertain parameter. We assume that the state equation $C(y, x, p) = 0$ uniquely defines y as an implicit function of x and p. The function $C : \mathbb{R}^{n_y + n_x + n_p} \mapsto \mathbb{R}^{n_y}$ is assumed to be twice continuously differentiable with invertible jacobian $\frac{\partial}{\partial y}C$. We also require the function $G = (g_1, \cdots, g_{n_g})^T : \mathbb{R}^{n_y + n_x + n_p} \mapsto \mathbb{R}^{n_g}$, defining the inequality constraints in the presented problem, as well as the objective $J(y, x, p) \in \mathbb{R}$ to be sufficiently smooth. When addressing uncertainty it is a common assumption to have some prior knowledge about the parameter p in such a way that it is restricted to remain in a known uncertainty set $\mathcal{U}_p$. In order to eventually get computationally tractable problems, this set must not be chosen too complicated. The usually considered uncertainty sets are polytopes or generalized balls. For this paper, we assume that the parameter p is restricted to an ellipsoid around a known nominal value $\bar{p}$. This means the uncertainty set is supposed to have the following representation

$$\mathcal{U}_p := \{p \in \mathbb{R}^{n_p} : p = \bar{p} + Q \cdot \delta,\ \|\delta\|_2 \leq 1\} = \{p \in \mathbb{R}^{n_p} : \|Q^{-1}(p - \bar{p})\|_2 \leq 1\}, \tag{2}$$

where $Q \in \mathbb{R}^{n_p, n_p}$ is an invertible matrix. Hence, the matrix $B := Q^{-T}Q^{-1}$ is symmetric positive definite and for the euclidean norm $\|\cdot\|_2$ we can define the related B-norm by $\|x\|_B := \sqrt{x^T B x}$. Consequently, $\|p - \bar{p}\|_B = \|Q^{-1}(p - \bar{p})\|_2$ holds and thus the uncertainty set can be written as

$$\mathcal{U}_p = \{p \in \mathbb{R}^{n_p} : \|p - \bar{p}\|_B \leq 1\}. \tag{3}$$

It is worth mentioning that the first order approximation approach proposed in [11] allows for more general uncertainty sets, namely generalized balls. There is no restriction to the euclidean norm when defining $\mathcal{U}_p$ in (2), but any Hölder q-norm $\|\cdot\|_q$ for $1 \leq q \leq \infty$ can be used. However, for the second order approximation approach it is essential to use the euclidean norm to attain a problem that is efficiently solvable.

Formulation of the robust counterpart. To incorporate the presence of uncertainty in the parameter p into the optimization problem, we use the classical worst-case approach. This has been proposed by Ben-Tal and Nemirovski, e.g. [1, 6, 14], and El Ghaoui et al. [7, 8] for conic optimization problems and has been adjusted to general nonlinear optimization problems with state equations by Diehl et al. [11] and independently by Zhang [12]. Following the notation of Diehl et al. we define

$$\phi_0(x) := \max_{y \in \mathbb{R}^{n_y}, p \in \mathbb{R}^{n_p}} J(y, x, p) \quad \text{s.t.} \quad C(y, x, p) = 0, \quad \|(p - \bar{p})\|_B \leq 1,$$

$$\phi_i(x) := \max_{y \in \mathbb{R}^{n_y}, p \in \mathbb{R}^{n_p}} g_i(y, x, p) \quad \text{s.t.} \quad C(y, x, p) = 0, \quad \|(p - \bar{p})\|_B \leq 1, \quad i = 1, \ldots, n_g.$$

Using these definitions we formulate the following worst-case version of the nominal problem, which is usually referred to as the ``robust counterpart'' of (1), cf. [14].

$$\min_{x \in \mathbb{R}^{n_x}} \phi_0(x) \quad \text{s.t.} \quad \phi_i(x) \leq 0 \quad i = 1, \ldots, n_g. \tag{4}$$

The construction of the robust counterpart guarantees that every x which is feasible for (4) is a feasible solution for the nominal problem (1) for any possible realisation of the parameter p from the uncertainty set $\mathcal{U}_p$. Further, every x which is optimal for (4), is a robust optimal solution to (1) in the sense that it provides the best worst-case performance over all possible p.

3. Approximation of the robust counterpart

Because of its nested structure the robust counterpart is computationally not tractable for general nonlinear functions. In [10] Ben-Tal, Nemirovski, and El Ghaoui provide an extensive discussion for which types of nominal problems the associated robust counterpart can be formulated as or at least can be approximated by a tractable problem. However, they do not consider the case of general nonlinear optimization problems with state equation. In [11] Diehl et al. adopt the idea of replacing the robust counterpart (4) by a tractable approximation

$$\min_{x \in \mathbb{R}^{n_x}} \tilde{\phi}_0(x) \quad \text{s.t.} \quad \tilde{\phi}_i(x) \leq 0 \quad i = 1, \ldots, n_g, \tag{5}$$

where the approximations $\tilde{\phi}_0(x)$ and $\tilde{\phi}_i(x)$ can be calculated more easily than the original worst-case functions $\phi_0(x)$, respectively $\phi_i(x)$.

First order approximation of the robust counterpart. Diehl et al. propose to replace in (4) the worst-case functions ϕ_0, ϕ_i as well as the state equation $C(y, x, p) = 0$ by their first order Taylor expansions. For given design variable x the linearization is carried out around the point $(\bar{y}, x, \bar{p})$ satisfying $C(\bar{y}, x, \bar{p}) = 0$, with $\bar{p}$ being the nominal value of the parameter p. By our assumptions from the very beginning of this paper, we are guaranteed the existence of such a point $\bar{y} = y(x, \bar{p})$. This leads to the following optimization problems

$$\tilde{\phi}_i(x) := \max_{\Delta y \in \mathbb{R}^{n_y}, \Delta p \in \mathbb{R}^{n_p}} g_i(\bar{y}, x, \bar{p}) + \partial_{(y,p)} g_i(\bar{y}, x, \bar{p}) \cdot \begin{pmatrix} \Delta y \\ \Delta p \end{pmatrix}$$

$$\text{s.t.} \quad \underbrace{C(\bar{y}, x, \bar{p})}_{=0} + \partial_{(y,p)} C(\bar{y}, x, \bar{p}) \cdot \begin{pmatrix} \Delta y \\ \Delta p \end{pmatrix} = 0, \quad \|\Delta p\|_B \leq 1, \tag{6}$$

where, for notational convenience, we used the following abbreviations for partial derivatives $\partial_p := \frac{\partial}{\partial p}$, $\partial_{(y,p)} := (\partial_y, \partial_p)$, and $\partial_{yp} := \frac{\partial^2}{\partial y \partial p}$. We remark that the approximation $\tilde{\phi}_0(x)$ of $\phi_0(x)$ is obtained analogously through replacing the function g_i by J. Thus, from now on we will restrict our explanations to $\tilde{\phi}_i$, keeping in mind that we just need to identify g_0 with J.

The optimization problems (6) are convex and can easily be solved analytically by eliminating Δy and then applying Hölder's inequality. The corresponding optimal value is

$$\tilde{\phi}_i(x) = g_i(\bar{y}, x, \bar{p}) + \left\| \left[(\partial_p g_i - \partial_y g_i (\partial_y C)^{-1} \partial_p C)(\bar{y}, x, \bar{p}) \right] Q \right\|_2.$$

Introducing adjoint sensitivities λ, λ_i one arrives at the following numerically efficient formulation of the ``approximated robust counterpart'' (5), cf.[11]:

$$\min_{x \in \mathbb{R}^{n_x}, \bar{y} \in \mathbb{R}^{n_y}, \lambda \in \mathbb{R}^{n_y}, \lambda_i \in \mathbb{R}^{n_y}} J(\bar{y}, x, \bar{p}) + \left\| \left[\partial_p J(\bar{y}, x, \bar{p}) + \lambda^T \partial_p C(\bar{y}, x, \bar{p}) \right] Q \right\|_2$$

$$\text{s.t.} \quad g_i(\bar{y}, x, \bar{p}) + \left\| \left[\partial_p g_i(\bar{y}, x, \bar{p}) + \lambda_i^T \partial_p C(\bar{y}, x, \bar{p}) \right] Q \right\|_2 \le 0, \quad i = 1, \dots, n_g,$$

$$C(\bar{y}, x, \bar{p}) = 0, \quad (7)$$

$$(\partial_y C(\bar{y}, x, \bar{p}))^T \lambda + (\partial_y J(\bar{y}, x, \bar{p}))^T = 0,$$

$$(\partial_y C(\bar{y}, x, \bar{p}))^T \lambda_i + (\partial_y g_i(\bar{y}, x, \bar{p}))^T = 0, \quad i = 1, \dots, n_g.$$

By using this adjoint formulation it is possible to avoid the explicit appearance of the inverse $(\partial_y C)^{-1}$ such that sparsity can be preserved. For more details cf. [11].

Second order approximation of the robust counterpart. Using only a first order approximation of the objective function as well as of the inequality constraints might not suffice to describe the effects of the uncertain parameter accurately enough. Avoiding this problem by choosing a smaller uncertainty set $\mathcal{U}_p$ will not be an option in most real-world applications. Hence, in this paper we adopt and continue the approach of Diehl et al. We also replace the state equation C with its first order Taylor expansion, but propose to approximate the objective function J as well as the inequality constraints g_i by their second order Taylor expansions. In this way we get the following approximations of the worst-case functions $\phi_i(x)$:

$$\tilde{\phi}_i(x) := \max_{\Delta y \in \mathbb{R}^{n_y}, \Delta p \in \mathbb{R}^{n_p}} g_i(\bar{y}, x, \bar{p}) + \partial_{(y,p)} g_i(\bar{y}, x, \bar{p}) \cdot \begin{pmatrix} \Delta y \\ \Delta p \end{pmatrix}$$

$$+ \frac{1}{2} (\Delta y^T, \Delta p^T) \left[\begin{pmatrix} \partial_{yy} g_i & \partial_{yp} g_i \\ \partial_{py} g_i & \partial_{pp} g_i \end{pmatrix} (\bar{y}, x, \bar{p}) \right] \begin{pmatrix} \Delta y \\ \Delta p \end{pmatrix} \quad (8)$$

$$\text{s.t.} \quad \underbrace{C(\bar{y}, x, \bar{p})}_{=0} + \partial_{(y,p)} C(\bar{y}, x, \bar{p}) \cdot \begin{pmatrix} \Delta y \\ \Delta p \end{pmatrix} = 0, \quad \|\Delta p\|_B \le 1. \quad (9)$$

As before, the linearized state equation in (9) yields $\Delta y = -(\partial_y C(\bar{y}, x, \bar{p}))^{-1} \cdot \partial_p C(\bar{y}, x, \bar{p}) \cdot \Delta p$ and thereby enables us to eliminate Δy from the above optimization problem. Introducing the notation

$$d_i(\bar{y}, x, \bar{p})^T := \left(\partial_p g_i - \partial_y g_i (\partial_y C)^{-1} \partial_p C \right)(\bar{y}, x, \bar{p}) \quad (10)$$

$$M_i(\bar{y}, x, \bar{p}) := \left((\partial_p C)^T (\partial_y C)^{-T} \partial_{yy} g_i (\partial_y C)^{-1} \partial_p C - \partial_{py} g_i (\partial_y C)^{-1} \partial_p C \right.$$

$$\left. - (\partial_p C)^T (\partial_y C)^{-T} \partial_{yp} g_i + \partial_{pp} g_i \right)(\bar{y}, x, \bar{p}) \quad (11)$$

we obtain for (8)-(9) the equivalent formulation

$$\tilde{\phi}_i(x) = \max_{\Delta p \in \mathbb{R}^{n_p}} g_i(\bar{y}, x, \bar{p}) + d_i(\bar{y}, x, \bar{p})^T \Delta p + \frac{1}{2} \Delta p^T M_i(\bar{y}, x, \bar{p}) \Delta p \quad \text{s.t.} \quad \|\Delta p\|_B \le 1. \quad (12)$$

Unlike in the case of first order approximations, it is now not possible to provide the optimal value $\tilde{\phi}_i(x)$ of problem (12) as closed analytic expression. However, an exact numerical solution of the above trust-region problem as well as the associated derivative $\tilde{\phi}'_i(x)$ can be calculated efficiently by means of a combined sensitivity and adjoint approach. This is an essential prerequisite for subsequently solving the approximation (12) of problem (4) using gradient-based optimization algorithms. To this end we recall some facts about trust-region problems.

4. The approximated worst-case functions formulated as trust-region problems

A generic form of a trust-region problem is

$$\max_{s \in \mathbb{R}^{n_s}} q(s; x) \quad \text{s.t.} \quad \|s\|_S \leq 1, \tag{13}$$

where the S-norm is defined as $\|s\|_S := \sqrt{s^T S s}$ for any symmetric positive definite matrix S, and the quadratic function $q(s; x)$ is defined as

$$q(s; x) := f(x) + c(x)^T s + \frac{1}{2} s^T H(x) s.$$

Here, $c(x) \in \mathbb{R}^{n_s}$ usually corresponds to the gradient $\nabla f(x)$, and $H(x) \in \mathbb{R}^{n_s,n_s}$ is a symmetric matrix that generally is chosen to be an approximation to the Hessian $\nabla^2 f(x)$.

First of all we note that the problem of interest (12) is a generic trust-region problem as described in (13). Since the matrices $\partial_{yy} g_i$ and $\partial_{pp} g_i$ are symmetric, and due to the fact that $(\partial_{py} g_i)^T = \partial_{yp} g_i$, the matrix $M_i(\bar{y}, x, \bar{p})$ is obviously symmetric. Furthermore, in the definition of the uncertainty set (3) matrix B was assumed to be symmetric positive definite.

Solution of the trust-region problem. To solve the presented trust-region problem we cite a slightly altered theorem from [15] concerning a characterization of the solution of (13).

Theorem 1. *Let $H(x) \in \mathbb{R}^{n_s,n_s}$ be an arbitrary symmetric matrix. Then the trust-region problem (13) has at least one global maximum. Furthermore, $\bar{s}$ is a global maximum of (13) if and only if there exists $\bar{\lambda} \geq 0$ such that the following three conditions hold*

$$\left(-H(x) + \bar{\lambda} S\right) \bar{s} = c(x), \tag{14}$$

$$\left(-H(x) + \bar{\lambda} S\right) \text{ is positive semidefinite}, \tag{15}$$

$$\text{either } \|\bar{s}\|_S = 1 \text{ or } \|\bar{s}\|_S < 1 \text{ and } \bar{\lambda} = 0. \tag{16}$$

The above conditions (14)-(16) can be used to compute an exact numerical solution of the trust-region problem (13). For this we have to distinguish two cases.
If $-H(x)$ is positive definite and the global maximum $s^{NA} := -H(x)^{-1} c(x)$ satisfies $\|s^{NA}\|_S \leq 1$, then $\bar{s} = s^{NA}$ is the solution of (13). This means that for fixed x we seek a solution $\bar{s}$ of the system of equations

$$k(s; x) := H(x) s + c(x) = 0, \quad \text{under the condition that } \|\bar{s}\|_S \leq 1. \tag{17}$$

Otherwise, i.e. if $-H(x)$ is not positive definite or if $\|s^{NA}\|_S > 1$ holds, one searches, again with x being fixed, for a solution $(\bar{s}, \bar{\lambda})$ of the system of equations

$$K(s, \lambda; x) := \begin{pmatrix} (-H(x) + \lambda S) s - c(x) \\ \frac{1}{\|s\|_S} - 1 \end{pmatrix} = 0, \quad \text{requiring } \bar{\lambda} \geq \max\{0, -\lambda_{min}(-H(x))\}. \tag{18}$$

Thereby $\lambda_{min}(-H(x))$ denotes the minimal eigenvalue of $-H(x)$.

Theorem 1 states that to any x there exists either a solution $\bar{s} = s(x)$ of (17) or a solution $(\bar{s}, \bar{\lambda}) = (s(x), \lambda(x))$ of (18) such that, in any of the two cases, $\bar{s}$ solves the considered trust-region problem (13). It is easy to observe that $\bar{\lambda}$ is always unique since we demanded that $\bar{\lambda} \geq \max\{0, -\lambda_{min}(-H(x))\}$ has to be fullfilled. On the other hand, $\bar{s}$ need not be unique. In particular, $\bar{s}$ is not unique if and only if $\bar{\lambda} = -\lambda_{min}(-H(x))$, and in addition $c(x)$ is orthogonal to the space of eigenvectors of $-H(x)$ to the eigenvalue $\lambda_{min}(-H(x))$. This is known as the ``hard case''. For details on how the system (18) can be solved efficiently, as well as for further information on trust-tregion techniques in general, we refer to [15].

5. The Solution and differentiation of the approximated worst-case functions

We now have everything at hand to determine the optimal value $\tilde{\phi}_i(x)$ of the central problem (12) as well as its derivative $\tilde{\phi}'_i(x)$. We have already seen that the approximated worst-case function (12) is a trust-region problem of the form

$$\tilde{\phi}_i(x) = \max_{s \in \mathbb{R}^{n_s}} q_{\tilde{\phi}_i}(s; x) \quad \text{s.t.} \quad \|s\|_S \leq 1, \tag{19}$$

where the quadratic function $q_{\tilde{\phi}_i}(s; x)$ is defined as

$$q_{\tilde{\phi}_i}(s; x) := g_i(\bar{y}, x, \bar{p}) + d_i(\bar{y}, x, \bar{p})^T s + \frac{1}{2} s^T M_i(\bar{y}, x, \bar{p}) s. \tag{20}$$

We consider the general case, when $-M_i(\bar{y}, x, \bar{p})$ is not positive definite, or when $\|s^{NA}\|_S \leq 1$ does not hold. Then, according to section 4, the following system of equations is employed to calculate a solution $\bar{s}_i$ of problem (19):

$$K_{\tilde{\phi}_i}(s, \lambda; x) := \begin{pmatrix} (-M_i(\bar{y}, x, \bar{p}) + \lambda S) s - d_i(\bar{y}, x, \bar{p}) \\ \frac{1}{\|s\|_S} - 1 \end{pmatrix} = 0. \tag{21}$$

As explained before, we search for a point $(\bar{s}_i, \bar{\lambda}_i) = (s_i(x), \lambda_i(x))$ with $K_{\tilde{\phi}_i}(s_i(x), \lambda_i(x); x) = 0$ such that $\bar{\lambda}_i \geq \max\{0, -\lambda_{min}(-M_i(\bar{y}, x, \bar{p}))\}$ holds. We remark that if $-M_i(\bar{y}, x, \bar{p})$ is already positive definite and $\|s_i^{NA}\|_S \leq 1$ is satisfied, all calculations can be carried out analogously by just replacing $K_{\tilde{\phi}_i}(s, \lambda; x)$ with $k_{\tilde{\phi}_i}(s; x)$. For the optimal value of (19) we can thus write

$$\tilde{\phi}_i(x) = q_{\tilde{\phi}_i}(\bar{s}_i; x) = q_{\tilde{\phi}_i}(s_i(x); x).$$

As already mentioned above, we also need to determine the derivative $\tilde{\phi}'_i(x)$ for being able to employ gradient-based algorithms to solve the approximated robust counterpart (12). Therefore, we have to provide a way to differentiate the presented trust-region problem (19) with respect to changes in x. Formally the desired derivative is given by

$$\tilde{\phi}'_i(x) = \partial_s q_{\tilde{\phi}_i}(\bar{s}_i; x) \cdot s'_i(x) + \partial_x q_{\tilde{\phi}_i}(\bar{s}_i; x).$$

Because it is too expensive to explicitly compute the derivative $s'_i(x)$ of the implicitly given function $s_i(x)$, we apply adjoint-techniques for computing $\tilde{\phi}'_i(x)$. Since by definition of $\bar{s}_i$ and $\bar{\lambda}_i$ the equation $K_{\tilde{\phi}_i}(\bar{s}_i, \bar{\lambda}_i; x) = 0$ is fulfilled, the following equality

$$\tilde{\phi}_i(x) = \tilde{\phi}_i(x) + \mu_i^T \cdot K_{\tilde{\phi}_i}(\bar{s}_i, \bar{\lambda}_i; x) = q_{\tilde{\phi}_i}(\bar{s}_i; x) + \mu_i^T \cdot K_{\tilde{\phi}_i}(\bar{s}_i, \bar{\lambda}_i; x). \tag{22}$$

holds for arbitrary $\mu_i \in \mathbb{R}^{n_s+1}$. Hence, the required derivative can be written as

$$\tilde{\phi}'_i(x) = \partial_s q_{\tilde{\phi}_i}(\bar{s}_i; x) \cdot s'_i(x) + \partial_x q_{\tilde{\phi}_i}(\bar{s}_i; x)$$
$$+ \mu_i^T \cdot \left(\partial_s K_{\tilde{\phi}_i}(\bar{s}_i, \bar{\lambda}_i; x) \cdot s'_i(x) + \partial_\lambda K_{\tilde{\phi}_i}(\bar{s}_i, \bar{\lambda}_i; x) \cdot \lambda'_i(x) + \partial_x K_{\tilde{\phi}_i}(\bar{s}_i, \bar{\lambda}_i; x) \right) \tag{23}$$
$$= \left(\partial_s q_{\tilde{\phi}_i}(\bar{s}_i; x) + \mu_i^T \cdot \partial_s K_{\tilde{\phi}_i}(\bar{s}_i, \bar{\lambda}_i; x) \right) \cdot s'_i(x) + \mu_i^T \cdot \partial_\lambda K_{\tilde{\phi}_i}(\bar{s}_i, \bar{\lambda}_i; x) \cdot \lambda'_i(x)$$
$$+ \partial_x q_{\tilde{\phi}_i}(\bar{s}_i; x) + \mu_i^T \cdot \partial_x K_{\tilde{\phi}_i}(\bar{s}_i, \bar{\lambda}_i; x) \tag{24}$$

We now choose μ_i as the solution of the system of equations

$$\begin{pmatrix} \partial_s K_{\tilde{\phi}_i}(\bar{s}_i, \bar{\lambda}_i; x)^T \\ \partial_\lambda K_{\tilde{\phi}_i}(\bar{s}_i, \bar{\lambda}_i; x)^T \end{pmatrix} \cdot \mu_i = \begin{pmatrix} -\partial_s q_{\tilde{\phi}_i}(\bar{s}; x)^T \\ 0 \end{pmatrix}. \tag{25}$$

That way one avoids having to compute $s_i'(x)$ and the desired derivative can be written as

$$\tilde{\phi}_i'(x) = \partial_x q_{\tilde{\phi}_i}(\bar{s}_i; x) + \mu_i^T \cdot \partial_x K_{\tilde{\phi}_i}(\bar{s}_i, \bar{\lambda}_i; x). \tag{26}$$

Note that the computation of the partial derivatives $\partial_s q_{\tilde{\phi}_i}$, $\partial_s K_{\tilde{\phi}_i}$, and $\partial_\lambda K_{\tilde{\phi}_i}$, required in (25), is trivial, cf. the definitions of $q_{\tilde{\phi}_i}$ and $K_{\tilde{\phi}_i}$ in (20) and (21). The calculation of the more complicated derivatives $\partial_x q_{\tilde{\phi}_i}$ and $\partial_x K_{\tilde{\phi}_i}$, needed in (26), will be discussed in the following subsection.

We remark that there might be points x where $\tilde{\phi}_i'(x)$ is nondifferentiable. This can occur in the aforementioned ``hard case'', that is, if (19) does not have a unique solution $\bar{s}_i$. To circumvent this difficulty we are currently working on a modification of the presented approach which uses similar computations.

Numerically efficient formulation. The calculation of the optimal value $\tilde{\phi}_i(x)$ requires that we explicitly compute the values $g_i(\bar{y}, x, \bar{p})$, $d_i(\bar{y}, x, \bar{p})$, and $M_i(\bar{y}, x, \bar{p})$. Looking back at the definition (10) and (11) of d_i and M_i, we note the appearance of the inverse $(\partial_y C)^{-1}$. To avoid the explicit calculation of $(\partial_y C)^{-1}$, we introduce so called ``direct sensitivities'', cf. [11].

$$\bar{l} := l(\bar{y}, x, \bar{p}) := (\partial_y C(\bar{y}, x, \bar{p}))^{-1} (\partial_p C(\bar{y}, x, \bar{p})) \,.$$

Thus, from (10) and (11) we obtain

$$d_i(\bar{y}, x, \bar{p})^T = (\partial_p g_i - \partial_y g_i \cdot l)(\bar{y}, x, \bar{p}) =: \tilde{d}_i(l(\bar{y}, x, \bar{p}), \bar{y}, x, \bar{p})^T = \tilde{d}_i(\bar{l}, \bar{y}, x, \bar{p})^T, \tag{27}$$

$$M_i(\bar{y}, x, \bar{p}) = \left(l^T \cdot (\partial_{yy} g_i) \cdot l - (\partial_{py} g_i) \cdot l - l^T \cdot (\partial_{yp} g_i) + \partial_{pp} g_i \right)(\bar{y}, x, \bar{p})$$
$$=: \tilde{M}_i(l(\bar{y}, x, \bar{p}), \bar{y}, x, \bar{p}) = \tilde{M}_i(\bar{l}, \bar{y}, x, \bar{p}). \tag{28}$$

Hence, direct sensitivities are additional variables that have to satisfy the supplemental equation

$$L(l, \bar{y}, x, \bar{p}) := (\partial_y C(\bar{y}, x, \bar{p})) l - (\partial_p C(\bar{y}, x, \bar{p})) = 0. \tag{29}$$

We note that $l \in \mathbb{R}^{n_y, n_p}$. Further, equation (29) is a matrix equation in $\mathbb{R}^{n_y, n_p}$ that implicitly defines the function $l(\bar{y}, x, \bar{p})$, which provides for all $(\bar{y}, x)$ the value $\bar{l} = l(\bar{y}, x, \bar{p})$ such that $L(\bar{l}, \bar{y}, x, \bar{p}) = 0$. Hence, formulation (27)-(28) enables us to efficiently compute $d_i(\bar{y}, x, \bar{p})$ and $M_i(\bar{y}, x, \bar{p})$. Moreover, representation (27)-(28) also helps to calculate the derivatives

$$\frac{d}{dx}(b^T M_i(\bar{y}, x, \bar{p}) \bar{s}_i), \quad \frac{d}{dx}(d_i(\bar{y}, x, \bar{p})^T b, \text{ and } \quad \frac{d}{dx} g_i(\bar{y}, x, \bar{p}), \tag{30}$$

where $b \in \{\bar{s}_i, \mu_i\}$. According to (20) and (21), these derivatives are needed to determine the partial derivatives $\partial_x q_{\tilde{\phi}_i}$, $\partial_x K_{\tilde{\phi}_i}$, which in turn are necessary to calculate $\tilde{\phi}_i'(x)$, cf. (26).

In order to circumvent the explicit computation of $l'(x)$ and $y'(x)$ when determining the total derivatives given in (30), we combine the described direct sensitivity approach with an adjoint approach that is similar to the one used in (22)-(26). This means, we have

$$g_i(\bar{y}, x, \bar{p}) = g_i(\bar{y}, x, \bar{p}) + \vartheta_0^T C(\bar{y}, x, \bar{p}),$$
$$d_i(\bar{y}, x, \bar{p})^T b = \tilde{d}_i(\bar{l}, \bar{y}, x, \bar{p})^T b + \vartheta_1^T L(\bar{l}, \bar{y}, x, \bar{p}) b + \vartheta_2^T C(\bar{y}, x, \bar{p}),$$
$$b^T M_i(\bar{y}, x, \bar{p}) \bar{s}_i = b^T \tilde{M}_i(\bar{l}, \bar{y}, x, \bar{p}) \bar{s}_i + \vartheta_3^T L(\bar{l}, \bar{y}, x, \bar{p}) \bar{s}_i + \vartheta_4^T L(\bar{l}, \bar{y}, x, \bar{p}) b + \vartheta_5^T C(\bar{y}, x, \bar{p}),$$

for arbitrary $\vartheta_0, \ldots, \vartheta_5 \in \mathbb{R}^{n_y}$. Proceeding similarly to (23)-(25), we choose $\vartheta_0, \ldots, \vartheta_5$ such that $l'(x)$ and $y'(x)$ need not be calculated. Hence, the desired derivatives can be written as

$$\frac{d}{dx} g_i(\bar{y}, x, \bar{p}) = \partial_x g_i(\bar{y}, x, \bar{p}) + \vartheta_0^T \cdot \partial_x C(\bar{y}, x, \bar{p}),$$

$$\frac{d}{dx} \left(d_i(\bar{y}, x, \bar{p})^T b \right) = \partial_x \left(\tilde{d}_i(\bar{l}, \bar{y}, x, \bar{p})^T b \right) + \partial_x \left(\vartheta_1^T L(\bar{l}, \bar{y}, x, \bar{p}) b \right) + \partial_x \left(\vartheta_2^T C(\bar{y}, x, \bar{p}) \right),$$

$$\frac{d}{dx} \left(b^T M_i(\bar{y}, x, \bar{p}) \bar{s}_i \right) = \partial_x \left(b^T \tilde{M}_i(\bar{l}, \bar{y}, x, \bar{p}) \bar{s}_i \right) + \partial_x \left(\vartheta_3^T L(\bar{l}, \bar{y}, x, \bar{p}) \bar{s}_i \right) + \partial_x \left(\vartheta_4^T L(\bar{l}, \bar{y}, x, \bar{p}) b \right)$$
$$+ \partial_x \left(\vartheta_5^T C(\bar{y}, x, \bar{p}) \right).$$

We remark that for finding the proper values of ϑ_j, one has to solve a sparse system of linear equations of dimension n_y for each ϑ_j, where the linear systems comprise the partial derivatives $\partial_l(\tilde{d}_i^T b), \partial_y(\tilde{d}_i^T b), \partial_l(b^T \tilde{M}_i \bar{s}_i), \partial_y(b^T \tilde{M}_i \bar{s}_i), \partial_l(Lb), \partial_y(Lb)$, and $\partial_y C$.

6. Application

To illustrate the presented second order approximation technique for the robust counterpart of a general uncertain nonlinear program with state equation we consider a generic example from shape optimization in structural mechanics. We optimize the geometry of a two-dimensional rod with two holes (see Figure 1), whose lower boundary is assumed to be fixed, hence being a homogeneous Dirichlet boundary. We suppose a uniform but uncertain loading $g \in \mathbb{R}^2$ at the upper boundary of the rod and a vanishing volume load f.

Since we want to optimize the geometry of the regarded structure, its shape $\Omega(x)$ is not fixed but changes depending on the design variables $x \in \mathbb{R}^{n_x}$. To this end, the boundary of $\Omega(x)$ is parameterized via piecewise cubic Bézier curves that depend on x. We also impose certain constraints on the geometry of $\Omega(x)$. First, we require that the Neumann boundary, that is the part of the boundary at which the surface force g is applied, has fixed length and is parallel to the Dirichlet boundary. Second, we introduce an upper bound on the volume, as well as a lower bound on the length of the rod. Third, we require a minimal diameter for the two holes. These geometry constraints can be written as $G(x) \leq 0$ with a smooth function $G : \mathbb{R}^{n_x} \mapsto \mathbb{R}^{n_G}$.

To model the mechanical behavior of the rod we introduce the two-dimensional linear elasticity equations as a PDE constraint. For the numerical treatment of the elasticity equations its weak formulation is discretized by a linear finite element approximation using the pure displacement approach. In particular, the state equation for our example can be written as

$$C(y, x, g) = A(x)y - M(x)f - N(x)g = 0, \tag{31}$$

where $A(x)$ denotes the stiffness matrix, $M(x)$ stands for the mass matrix, and $N(x)$ is the mass matrix for the boundary of the structure $\Omega(x)$. The state variable $y \in \mathbb{R}^{n_y}$ describes the displacement of the structure under the acting load. Since we assumed a vanishing volume load, the term $M(x)f$ in (31) disappears in our example. For Young's modulus and Poisson's ratio that decisively influence the stiffness matrix, we choose $E = 210$ GPa and $\nu = \frac{1}{3}$.

For the objective function that is to be minimized, we consider the mean square norm of the displacement of the structure:

$$J(y, x) := \frac{\|y\|_2^2}{\text{vol}(\Omega(x))}.$$

Now, our application example can be written in the general form (1) introduced in section 2:

$$\min_{y \in \mathbb{R}^{n_y}, x \in \mathbb{R}^{n_x}} J(y, x) \quad \text{s.t.} \quad G(x) \leq 0, \ C(y, x, p) = 0, \tag{32}$$

where the acting load $g =: p$ is assumed to be the uncertain parameter. Note that in the presented example neither the objective function nor the inequality constraints depend directly on the uncertain parameter. However, since the state variable y implicitly depends on g via the state equation (31), the objective function J is implicitly influenced by changes of g. As motivated in sections 1 and 2, we assume that the exact value of g is not known. Instead we suppose that the actual load results from a nominal load $\bar{g} \in \mathbb{R}^2$ which is disturbed by an arbitrary force of magnitude less than 5% of $\|\bar{g}\|_2$. Here, $\bar{g}$ is chosen to act in axial direction of the rod. Hence, we define the uncertainty set, to which we assume the load g is restricted, as

$$\mathcal{U}_g := \{g \in \mathbb{R}^2 : g = \bar{g} + Q \cdot \delta, \|\delta\|_2 \leq 1\},$$

with $Q := 0.05 \cdot \|\bar{g}\|_2 \cdot I$, where $I \in \mathbb{R}^{2,2}$ denotes the identity matrix.

Figure 1 shows the different results of the non-robust optimization and the robust optimization. For the non-robust case the uncertainty in the parameter g is ignored and problem (32) is solved for $g = \bar{g}$, obtaining a nominal solution x^*. For the robust case, however, we solve the second order approximation of the robust counterpart associated with (32) to receive a robust solution x_R^*. In both cases, the corresponding optimization problem is solved using a standard SQP-method. Comparing the worst-case objective values of x^* and x_R^* we note that the robust solution has a 3.9 % better worst-case performance than the nominal solution. This improvement comes at the price of a nominal performance which is 2.2 % worse than the one of the nominal solution. Of course, the ratio of the gained robustness and the lost nominal performance is not as significant as for the examples presented in [1, 2, 14]. Nevertheless the results show that the robust optimization paradigm can efficiently be applied to simulation based optimization problems involving a PDE as state equation. Also, we expect more significant results when considering slimmer structures. Further, we plan to incorporate uncertainty in the volume force f as well as in the Young's modulus E and the Poisson's ratio ν. In addition, we want to consider objective functions that depend directly on uncertain parameters like for example the compliance.

Fig. 1: On the left, the starting geometry before optimization is shown. In the middle, the optimal shape for the nominal loading is depicted. On the right, the approximated robust design, which incorporates uncertainty in the applied loading, is displayed. The colors show the norm of the displacement and use the same scale for all three cases.

7. Conclusion

We have presented a second order approximation for the robust counterpart of general uncertain nonlinear programs. Adopting the idea of Diehl et al. from [11], we showed how second order approximations of the worst-case functions can be formulated as trust-region problems, and how these can be solved and differentiated efficiently by employing a combined sensitivity and adjoint approach. We applied the presented techniques to a generic shape optimization problem from structural mechanics, with the acting surface force being an uncertain parameter.

Acknowledgements

We like to thank the Deutsche Forschungsgemeinschaft (DFG) for funding this project within the Collaborative Research Center (CRC) 805.

References

[1] A. Ben-Tal, A. Nemirovski. *Robust Truss Topology Design via Semidefinite Programming.* SIAM Journal on Optimization Vol. 7, pp. 991-1016, 1997.

[2] A. Ben-Tal, A. Nemirovski. *Robust solutions of linear programming problems contaminated with uncertain data.* Mathematical Programming, Vol. 88, pp. 411-421, 2000.

[3] J. R. Birge, F. Louveaux. *Introduction to Stochastic Programming.* Springer Series in Operations Research, Springer-Verlag, New York, 1997.

[4] A. Prékopa. *Stochastic Programming.* Kluwer Academic Publishers, Dordrecht, Boston, 1995.

[5] A. Shapiro, D. Dentcheva, A. Ruszczyński. *Lectures on Stochastic Programming: Modeling and Theory.* SIAM, Philadelphia, USA, 2009.

[6] A. Ben-Tal, A. Nemirovski. *Robust Convex Optimization.* Mathematics of Operations Research, Vol. 23, pp. 769-805, 1998.

[7] L. El Ghaoui, H. Lebret. *Robust Solutions to Least-Squares Problems with Uncertain Data.* SIAM Journal on Matrix Analysis and Applications Vol. 18, pp. 1035-1064, 1997.

[8] L. El Ghaoui, F. Oustry, H. Lebret. *Robust Solutions to Uncertain Semidefinite Programs.* SIAM Journal on Optimization Vol. 9, pp. 33-52, 1998.

[9] D. Bertsimas, D. B. Brown, C. Caramanis. *Theory and applications of Robust Optimization.* Submitted for publication, 2010. Available at arXiv:1010.5445v1.

[10] A. Ben-Tal, A. Nemirovski, L. El Ghaoui. *Robust Optimization.* Princeton University Press, 2009.

[11] M. Diehl, H.G. Bock, E. Kostina. *An approximation technique for robust nonlinear optimization.* Mathematical Programming, Vol. 107, pp. 213-230, 2006.

[12] Y. Zhang. *A General Robust-Optimization Formulation for Nonlinear Programming.* Journal of Optimization Theory and Applications, Vol. 132, pp. 111-124, 2007.

[13] D. Bertsimas, O. Nohadani, K. M. Teo. *Nonconvex Robust Optimization for Problems with Constraints.* INFORMS Journal on Computing Vol. 22, pp. 44-58, 2010.

[14] A. Ben-Tal, A. Nemirovski. *Robust optimization - methodology and applications.* Mathematical Programming, Vol. 92, pp. 453-480, 2002.

[15] A. R. Conn, N. I. M. Gould, Ph. L. Toint. *Trust-Region Methods.* SIAM, Philadelphia, 2000.

Approaches for assessment of non-determinism in different stages of engineering product design using finite element analysis

Dirk Vandepitte[1,a] and David Moens[2,b]

[1]K.U.Leuven -- PMA division, Celestijnenlaan 300-B -- box 2420, 3001 Leuven, Belgium

[2]Lessius Hogeschool, De Nayer Institute, Department of Applied Engineering,
K.U.Leuven Association, J. De Nayerlaan 5, 2860 Sint-Katelijne-Waver, Belgium

[a]dirk.vandepitte@mech.kuleuven.be, [b]david.moens@mech.kuleuven.be

Keywords: probability density function, interval number, fuzzy number, design criteria, finite element analysis

Abstract. Development of a complicated technical problem encompasses different successive decisions that are based on techical analysis and engineering judgment. In each of these steps, some degree of non-determinism is inevitable. The paper discusses different types of non-deterministic parameters that may be relevant in different stages of engineering analysis. The entire cycle of development of a product is considered, and it is shown that the relevance of methods is different in different stages of the development cycle.

A classification of different types of non-deterministic properties is presented. Based on the nature of these different classes of model properties, it is discussed to what degree each of these fits in the framework of either a probabilistic or a non-probabilistic concept.

The availability of realistic data in an appropriate format is another issue that should be taken into account. A validated probabilistic representation is usually only possible after an extensive campaign of data acquisition has been conducted, or at least after sufficient data have been collected to allow for a reliable estimation of a statistical model. A study of scientific literature shows that validated information is not always available.

A general conclusion is that probabilistic methods are applicable in later stages of development, when a sufficiently large database of product data has been gathered. Probabilistic approaches are perfectly suited for conditions when the product is already in service. Possibilistic analysis on the other hand is best suited for application in cases when the data set about the product at hand is still incomplete.

Introduction

Engineering design is the activity of design and development of technical products. A technical product is built to fulfil a well specified function under more or less well prescribed conditions of utilisation. The complexity of modern technical products tends to increase systematically, increasing the need for thorough design analysis. This process consists of a number of analysis verifications on a virtual product. A common procedure for design verification is finite element analysis, a numerical method for the simulation of the effect of mechanical or thermal loads on a product. As most product parameters are undetermined in the initial phases of design, a range of non-deterministic properties have to be taken into account. This paper discusses the effects of non-determinism on engineering analysis using the finite element method (FE).

There is a clear current trend to use the increasing computational power of modern computer hardware on uncertainty and variability, rather than on ever more detailed FE models. Probabilistic FE analysis already built up a long-standing tradition with increasing capabilities to analyse the effect of numerous uncertain and variable parameters in FE models. Examples can be found in [1], [2], [3], [4], [5], [6] and [7]. On the other hand, non-probabilistic methods have a much shorter history. Recently, a number of non-probabilistic approaches for non-deterministic analysis are emerging. The Interval FE analysis is based on the interval concept for the description of non-deterministic model

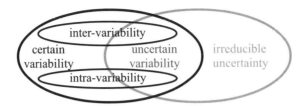

Fig. 1: Classes of variability and uncertainty and their overlaps

properties. The Fuzzy FE analysis is basically an extension of the IFE analysis. Examples include [8], [9], [10], [11], [12], [13], [14] and [15].

Definitions

In literature, the use of the terminology *error*, *uncertainty* and *variability* is not unambiguous. Different researchers apply the same terminology but the meaning attached to these is rather inconsistent. This necessitates a profound clarification of the terminology for each publication which treats uncertainties. This work does not propose a new terminology, but applies the terminology proposed by Oberkampf [16]. Some additional nuances are, however, necessary.

The term *variability* covers *the variation which is inherent to the modelled physical system or the environment under consideration*. Generally, this is described by a distributed quantity defined over a range of possible values. The exact value is known to be within this range, but it will vary from unit to unit or from time to time. Ideally, objective information on both the range and the likelihood of the quantity within this range is available. Some literature refers to this variability as *aleatory uncertainty* or *irreducible uncertainty*, referring to the fact that even when all information on the particular property is available, the quantity cannot be deterministically determined.

An *uncertainty* is *a potential deficiency in any phase or activity of the modelling process that is due to lack of knowledge*. The word *potential* stresses that the deficiency may or may not occur. This definition basically states that uncertainty is caused by incomplete information resulting from either vagueness, nonspecificity or dissonance [17]. Vagueness characterises information which is imprecisely defined, unclear or indistinct. It is typically the result of human opinion on unknown quantities (``the density of this material is around x''). Nonspecificity refers to the availability of a number of different models that describe the same phenomenon. The larger the number of alternatives, the larger the nonspecificity. Dissonance refers to the existence of conflicting evidence of the described phenomenon, for instance when there is evidence that a quantity belongs to disjoint sets. Possibly, limited objective information is available, for instance when a range of possible values is known. In most cases, however, information on uncertainties is subjective and based on some expert opinion. Others in literature refer to this uncertainty as *reducible*, *epistemic* or *subjective uncertainty*.

An *error* is defined as *a recognisable deficiency in any phase of modelling or simulation that is not due to lack of knowledge*. The fact that the error is recognisable states that it should be identifiable through examination, and as such is not caused by lack of knowledge. This means that the error could be avoided by an alternative approach which is known to be more accurate, but which is possibly limited in practical applicability by computational cost or other practical considerations. A further distinction between *acknowledged* and *unacknowledged* errors is possible. Errors will not be considered further in this paper.

These definitions are partially overlapping, as illustrated in Fig.1.

Assessment of non-deterministic parameters in the engineering process

Over the entire life of a technical product many sources of non-determinism may be relevant. This section describes briefly the process, and it identifies the phases when the uncertainties and variabilities play a role. The case of a product designed to withstand mechanical loads is taken as an example, but other cases are similar.

Overview of stages in the engineering process The engineering process typically consists of a number of successive phases :

1. definition of product specifications and design data, and definition of load cases

2. definition of a preliminary design and initial analysis of its feasibility

3. gradual design refinement and improvement, and specification of design details, concluded with the definition of final design

4. definition of the production process

5. production startup and quality control

6. operations of the product in service conditions

Each of these phases considers a number of inputs, some of which may be uncertain. Each of these phases is concluded with a decision.

Discussion of introduction of non-determinism in the engineering process Uncertainties and variabilities play a role in each of the phases that are listed. Throughout the discussion, the example of a truck chassis will be presented for the purpose of illustration.

1. **definition of product specifications** , design data and load cases : This phase includes the establishment of a list of input design data (product requirements) and conditions of utilisation of the product. These specifications are often rather general. For the purpose of technical analysis however, numerical data are required, preferably with a maximum degree of precision. It is common to include a margin of safety. However, this margin should not be too large, to avoid unnecessary overconservatism and uneconomical design. Several factors complicate the specification of precise data. Design data are still uncertain because some design parameters will be specified only during the subsequent process of design refinement. Specifications may be imprecise as several design variants of a product should be realised with a maximum degree of commonality. Component commonality is desired to reduce the number of components that are produced by a company and to simplify maintenance. On the other hand, commonality reduces the options for optimisation of a product.

   As far as technical design specifications are concerned, several requirements should be met : strength of the product, static and dynamic stiffness, fatigue life, ... The technical analysis that is required to verify these properties are very different, and further, these qualities depend on different product properties. Strength and fatigue life are the result of local design details, and as such, the data that are required to verify these properties are available only later in the analysis.

   The availability of design specifications depends very much on the industrial sector. In several areas design criteria are well established, based on many years of expertise with similar designs in previous cases. Design standards exist in the civil engineering sector, covering a wide spectrum of load cases in nominal and exceptional conditions, and these standards are even extensively documented. Other specific sectors of industry such as aircraft structures, pressure vessels, hoisting equipment also have well established design criteria. Standards are defined by independent normalisation and standardisation bodies, and insurance companies verify the correct application of standards before an insurance agreement is signed. In most other sectors of industry however, generally accepted standards are not available and each manufacturer has to decide for himself on the criteria to be applied. The determination of specifications is a rather delicate compromise between operational safety and economical design. Almost all consumer products are in this category.

In a limited number of industrial products, probability of failure is prescribed. An example is space industry, with a prescribed value on the reliability of launch vehicles. This value is however theoretical as the actual failure rate of launches does not match the prescribed numbers.

For the case of the truck, there are many product specifications, such as the type of load that the truck should transport, the maximum design load, its mission profile, the maximum dimensions, ... The mission profile may be very different : long haul for transcontinental transport on motorways, medium to short haul such as for concrete mixer trucks, very short range with very frequent stops and starts such as for garbage collection. The determination of loads is based to a large extent on experience with previous models. The load history can be measured by instrumentation of an existing vehicle. Typical mission profiles can be deduced, and used for later design development. Standards for trucks cover only part of the design requirements.

This phase in the design exercise should be concluded with a set of requirements that is as concise as possible, if relevant including statistical data.

2. **definition of preliminary design and initial analysis** After basic design requirements are formulated, one or more initial concepts are proposed for the newly designed product. Comparative design analysis typically uses so-called concept models that represent the global characteristics of the product without details. A correct focus on parameters that drive the design lay-out is crucial in this stage. For the sake of effective product design in subsequent phases, it is important that crucial design decisions are taken as early as possible. The pressure on design and development departments in companies to shorten product design cycles grows continuously. Unfortunately, complete product design data are usually not yet available at that stage, and data imprecisions have to be taken into account. Conservatism is absolutely required, yet without being excessive.

In the truck case, the concept model consists of discrete elements representing flexible components and discrete masses such as the engine and the fuel tank. The size and the filling percentage of the fuel tank (and sometimes also its position) being uncertain, the analysis has to take into account a relevant range of parameter settings.

Relatively few people are typically involved with this phase of design, and experience shows that most companies have only few experts who are qualified to define relevant inputs. The concept of subjective probability is therefor not applicable here. On the other hand, this design phase is concluded with a preliminary yet clear definition of the product concept. In the truck case, primary structural members can now be specified.

3. **design refinement, leading to final design** After the product concept is established, product design should be gradually refined. As design activities proceed --- often in separate design teams with different responsibilities --- more and more design data become available. Numerical models are refined, and detailed design analysis becomes feasible. The result set of the analysis grows likewise, and each output quantity is subject to verification of design criteria. Not only global criteria but also local criteria can now be verified, possibly including a safety factor. Each criterion is expressed as an inequality, and the degree by which it is fulfilled is unspecified and thus uncertain.

In the truck case detailed design includes the determination of details such as the position of holes and joints, the type of joints (welded, bolted, each with their inherent uncertainties), sizes of secondary structural members, ... The number of details is so large that it is sometimes impossible to include them all into a numerical model, inevitably increasing the uncertainty on the product behaviour. This statement is especially true for local design details that typically affect local product response, such as fatigue life.

This design phase is concluded with a concise complete set of design specifications, including all design details.

4. **production process definition** After nominal design parameters are specified, the entire process of production and assembly should be outlined. However, each step in the production process has its own range of accuracy that can be achieved. It depends on the quality of the specific production machine on which the component is manufactured and on the skills of the machine operator. The required accuracy is specified by the so-called geometrical tolerances. A tolerance is a min-max range and each measure should be verified to be within that range. The specification of a production process translates into the definition of a parameter range. However, on the production machine, nominal values are set.

For the truck case, nominal machine parameters have to be set on all machines for cutting, sawing, punching, drilling, milling, folding, grinding, ... This list contains precise, unique data, to be used at the production machinery of the manufacturer. A list of tolerances should be added specifying the ranges on geometrical properties.

5. **actual production and quality control** With the machine parameters set in the previous phase, actual production can be started. The result of production operations on each individual product is then subjected to some kind of quality control. However, even with a precise setting of machine parameters, the properties of each individual product are never identical. A quality control procedure is then required to verify if each product meets the standards. Different levels of quality control are used : no control on semi-finished products, implicit verification by obvious deficiencies, a predefined sampling procedure, full quality control on each individual product, extensive qualification and acceptance tests.

The results of quality control is interpreted in two different ways. At the level of each *individual* product, the verification of quality leads to either acceptance of the product or to rejection and scrapping. This decision is of a yes/no type. At the level of a *complete batch* of production, a quality distribution can be established expressing which percentage of products meets a desired quality level.

In some production facilities, non-conformancy procedures can be used, when a product exhibits an acknowledged deficiency, and the cost of scrapping is considered too high, it may be preferred to rework it and adapt the design to the observed deficiency. A new design analysis is required in that case. The result of this procedure is a yes/no decision.

6. **operations in service conditions** A well-designed product usually behaves well in normal service conditions, provided that the user or operator respects the conditions and limits of utilisation. However, unanticipated incidents may inflict damage on the product. If such an incident occurs, the operator usually verifies the extent of damage and he decides if the product should be repaired. Most mechanical systems further exhibit some kind of wear or damage accumulation (e.g. due to fatigue) over their economical lifetime. If the extent of wear or damage becomes such that nominal operation of the product gets dangerous or unreliable, it is common to replace the wear-sensitive or damaged components by new ones. This decision may be based on different criteria : preplanned after some fixed period of utilisation, by continuous monitoring of the performance of the product, or by more or less incidental observation of a deficiency. In the first or the second case, this decision may be prepared by previous expertise that is gathered after

careful examination of the operational performance of previous examples of a similar product. The decision to repair a component is a yes/no decision.

Unappropriate utilisation of the product may obviously lead to unforeseen events, which may have a negative influence on the appreciation of the user. This condition is not considered here.

Summary of decisions on non-determinism in successive steps of engineering analysis In all phases of engineering analysis of a technical component over its product lifetime, all decisions are crisp. They are either yes/no decisions or they involve the specification of precise values or a range of values. Availability of data is usually insufficient to support a statistical interpretation of design decisions. Generally speaking, the interval or the fuzzy concepts is most appropriate for technical analysis.

Statistical interpretation may be relevant in three out of the six stages that are listed above :

- the specification of design loads or the specification of a probability of failure may be based on a statistical interpretation

- product quality and control may lead to the establishment of quality distribution over the full range of nominally identical products

- the decision to replace a component during its service life may be based on statistical information gathered previously with similar products

Availability of data

Next to analysis methods, also numerical data must be available in the appropriate format to allow for a reliable analysis. The authors have conducted a review of journal publications in the field of non-deterministic analysis in structural dynamics. Only the numerical examples are considered, with a focus on how the input data are described. Most of the papers that listed are published in the journal for Computer Methods in Applied Mechanics and Engineering. Uncertainty mostly applies on material constitutive data, and for this reason several materials related journals are listed as well. The review is restricted to linear elastic material constants as they are the basic data that characterises the material constitutive behaviour in the most common condition of utilisation. These material data are most widely documented. Non-elastic and visco-elastic behaviour are left out of this study. Material data in these ranges are less available, and they also much more case-dependent.

Probabilistic models All considered case studies pertain to applications in structural dynamics and also in static structural analysis with random stiffness characteristics : the Integral spacecraft[1], a 12-storey building [2], a test coupon in Glare [3], the dynamics of a rod [4], a micro-electromechanical system [5], a truss structure [6] and a fibre-reinforced composite [7]. Only the first publication [1] gives a reference to data that are based on measurements or on thorough analysis. The authors of the other papers content themselves with assumptions on the nature and the quantification of stochasticity. Quite different levels are assumed for the coefficients of variation (ranging from 4% to 30%). Sometimes the coefficients of variation are different for different properties, and sometimes they are not. Sometimes a correlation between different properties is assumed, and sometimes properties are independent. The models for spatial variation are very diverse, with different assumptions for correlation length.

Non-probabilistic models Non-probabilistic models are used in different conditions, usually when limited data are available, and when a probabilistic interpretation is not required. The cases that are considered include a truck-trailer combination [8], a suspension triangle [9], an impactor [10], a cylindrical structure in a spacecraft [11], a vehicle windshield [12], [13], an aircraft structure [14] and the heat affected zone in a welded structure [15]. Non-probabilistic models are fed with input data that are only subjectively linked to realistic problem data. If the bounds of the interval are well defined, and if the non-probabilistic analysis procedure does not introduce artificial conservatism, and the output is a realistic set of bounds on output quantities.

Discussion of parameter data For uncertain variabilities, a representation by a single random quantity is generally not sufficient. Engineering scientist Freudenthal [18] stated in 1961 that "... *ignorance of the cause of variation does not make such variation random.*". By this, he means that when crucial information on a variability is missing, it is not good practice to model it as a probabilistic quantity represented by a single random PDF. On the contrary, in this case it is mandatory to apply a number of different probabilistic models to examine the effect of the chosen PDF on the result. For instance, when the range of the variability is known but the information on the likelihood is missing, all possible PDFs over the range should be taken into consideration in the analysis. The analyst will generally select only a few probabilistic models which he considers consistent with the limited available information or most appropriate to obtain as much knowledge as possible on the result. Another important criterion in the selection of the type of distribution is the nature of the distribution function itself and its relation to the phenomena that is represents. The risk function is a useful indicator in this respect.

Material data The mechanical properties of most common structural materials, especially metals and unreinforced polymers, are relatively well known. However, the range of materials is very wide, and properties may differ with precise chemical composition, with thermal treatment and they may even be different with different manufacturers. In addition, material properties have some scatter. However, over all physical and mechanical properties that a material exhibits, mass and stiffness usually are fairly close to their nominal values, unlike strength, which depends strongly on chemical composition and heat treatment. Thickness of the unworked piece also plays a role in material strength, with strength decreasing with increasing thickness.

Non-determinism in the properties of a specific material is a case of variability. Real materials are characterised in experimental measurements. The size of the set of measurements that are taken in identical conditions determines if a probability density function can be established with sufficient accuracy. If a sufficient number of individual measurements are available, the variability can be considered certain.

Metals and polymer materials databases The first source for metals and polymer materials data is a materials database. The MIL-handbook [19], which is now published on the web, and other web based databases such as `matweb` [20] and `efunda` [21] contain a large number of records for many material variants, even from different manufacturers. They usually specify nominal values, sometimes complemented with an indication of the probability distribution.

These databases do not give any indication on the spatial scatter within a test coupon. The test procedure implies that stiffness values are averaged numbers of the length of the sensor that is used, whereas strength is based on a local values in the section of fracture. Experimental data on spatial scatter are not available.

Other model properties In addition to material properties and geometrical dimensions and shapes, other FE model characteristics exhibit some kind of uncertainty or variability as well.

A delicate property is the boundary condition with which a structure is attached to the environment. Only one reference has been identified that addresses uncertainty on boundary conditions for buckling analysis of cylindrical shells with random boundary geometric imperfections [22]. FE models typically use either pinned of fixed conditions. In a pinned connection displacements are prescribed and rotations are free, and in a fixed connection both displacements and rotations are fixed. These conditions correspond to an infinitely stiff connection, which can never be realised in practice. The stiffness of the connection may be very small or very large, but it is always finite. The non-determinism has definitely a character of uncertainty, and an interval number or a fuzzy number seems to be the best representation.

Damping is another unknown quantity. Physically realistic models for damping are not available, and it may even be hard to characterise damping from experiments. An interval number is again the most appropriate model.

Observations on material data After bringing together all the information and material data that are reported in literature, the following general tendencies are observed :

- probabilistic methods provide more information than non-probabilistic methods; however, both families are highly complementary

- the number of publications on probabilistic methods exceeds the non-probabilistic ones

- almost all publications refer to aleatory uncertainty in material parameters, but there are very few references to uncertainty on other important FE model parameters that are not precisely known, such as boundary conditions

- very few publications refer to validated data, and most authors who publish in the leading scientific journal content themselves with assumptions on the non-deterministic nature of the model parameters

- very different values are assumed for the coefficients of variation on material parameters such as Young's modulus : from 4% to even 30% for isotropic materials

- literature does not provide any evidence on values for spatial scatter; correlation length is based on assumptions, apparently related to the length of the component

- correlation between model parameters is not taken into account

Conclusion

The emerging non-probabilistic approaches are redefining the landscape for non-deterministic FE analysis. It is the aim of this paper to give insight into the possible useful applications of these approaches, referring to the generally accepted and widely adopted probabilistic approach.

Different sources of uncertainty are reviewed, and it is concluded that the probabilistic approach remains the most interesting to tackle problems that are subject to complete and objective probabilistic influences. However, in the presence of uncertain quantities that require subjective information in order to be described numerically, the interval and fuzzy approach become increasingly interesting. Especially for uncertainties, the fuzzy concept is very appropriate because of its implicit subjective nature.

Researchers follow diverse strategies when they introduce non-determinism in their engineering analysis, and the type of data that are available does not necessarily match with the objectives of the analysis. The availability of data determines the type of non-deterministic analysis that can be executed without unintentional misrepresentation of data and inadvertent introduction of unvalidated assumptions. Inversely, a specific type of analysis can only be executed when the model data are available in a suitable format. The appropriate data format depends on the phase of development of structure that is considered and on the type of parameter that is modelled : material data, geometrical data, loads data, boundary conditions and spatial distribution of model parameters.

The authors perceive a need for a coordinated effort by the scientific research community to collect reliable data on different types of model parameters in an appropriate format for non-deterministic analysis and to make available these data to their fellow researchers and to the engineering community.

Acknowledgements

This work was funded by the Flemish regional agency for transfer of science to industry (IWT project 060043).

References

[1] Pellissetti, M., Capiez-Lernout, E., Pradlwarter, H., Soize, C., Schuëller, G.I., Reliability analysis of a satellite structure with a parametric and a non-parametric probabilistic model, *Computer Methods in Applied Mechanics and Engineering*, 2008, 198(2), pp. 344--357

[2] Schuëller, G.I., Efficient Monte Carlo simulation procedures in structural uncertainty and reliability analysis - recent advances, *Structural Engineering and Mechanics*, 2009, 32(1), pp. 1--20

[3] Chung, D.B., Gutierrez, M.A., de Borst, R., Object-oriented stochastic finite element analysis of fibre metal laminates, *Computer Methods in Applied Mechanics and Engineering*, 2005, 194(12-16), pp. 1427--1446

[4] Sarkar, A., Ghanem, R., Mid-frequency structural dynamics with parameter uncertainty, *Computer Methods in Applied Mechanics and Engineering*, 2002, 191(47-48), pp. 5499--5513

[5] Agarwal, N., Aluru, N.R., Stochastic modeling of coupled electromechanical interaction for uncertainty quantification in electrostatically actuated MEMS, *Computer Methods in Applied Mechanics and Engineering*, 2008, 197(43-44), pp. 3456--3471

[6] Falsone, G., Ferro, G., An exact solution for the static and dynamic analysis of FE discretized uncertain structures, *Computer Methods in Applied Mechanics and Engineering*, 2007, 196(21-24), pp. 2390--2400

[7] Stefanou, G., Papadrakakis, M., Stochastic finite element analysis of shells with combined random material and geometric properties, *Computer Methods in Applied Mechanics and Engineering*, 2004, 193(1-2), pp. 139--160

[8] Moens, D., Vandepitte, D., Interval sensitivity theory and its application to frequency response envelope analysis of uncertain structures, *Computer Methods in Applied Mechanics and Engineering*, 2007, 196(21-24), pp. 2486--2496

[9] Massa, F., Tison, T., Lallemand, B., A fuzzy procedure for the static design of imprecise structures, *Computer Methods in Applied Mechanics and Engineering*, 2006, 195(9-12), pp. 925--941

[10] Massa, F., Rufin, K., Tison, T., Lallemand, B., A complete method for efficient fuzzy modal analysis, *Journal of Sound and Vibration*, 2008, 304(1-2), pp. 63--85

[11] Moens, D., Vandepitte, D., Recent advances in non-probabilistic approaches for non-deterministic dynamic finite element analysis, *Archives of Computational Methods in Engineering*, 2006, 13(3), pp. 389--464

[12] Moens, D., De Munck, M., Vandepitte, D., Envelope frequency response function analysis of mechanical structures with uncertain modal damping characteristics, *Computer Methods in Engineering Science*, 2006, 22(2), pp. 129--149

[13] d'Ippolito, R., Tabak, U., De Munck, M., Donders, S., Moens, D., Vandepitte, D., Modelling of a vehicle windshield with realistic uncertainty, *Proceedings of ISMA 2006*, 2006, pp. 2023--2032

[14] Noor, A.K., Starnes, J.H.Jr., Peters, J.M., Uncertainty analysis of composite structures, *Computer Methods in Applied Mechanics and Engineering*, 2000, 185(2-4), pp. 413--432

[15] Khaled, A.-T., Noor, A.K., Uncertainty analysis of welding residual stress fields, *Computer Methods in Applied Mechanics and Engineering*, 1999, 179(3-4), pp. 327--344

[16] Oberkampf, W., DeLand, S., Rutherford, B., Diegert, K., and Alvin, K., A New Methodology for the Estimation of Total Uncertainty in Computational Simulation, *Proceedings of the 40th AIAA/ASME/ASCE/AHS/ASC Structures, Structural Dynamics and Materials Conference, AIAA-99-1612*, 1999, pp. 3061--3083

[17] Klir, G. and Folger, T., *Fuzzy Sets, Uncertainty and Information*, Prentice Hall, Englewood Cliffs, 1988.

[18] Freudenthal, A., Fatigue Sensitivity and Reliability of Mechanical Systems, especially Aircraft Structures, *WADD Technical Report 61-53*, 1961

[19] United States Department of Defense, Military Handbook -- metallic materials and elements for aerospace vehicle structures, edition 5J, 2003, http://www.weibull.com/knowledge/milhdbk.htm

[20] http://www.matweb.com

[21] http://www.efunda.com

[22] Schenck, C.A., Schuëller, G.I., Buckling analysis of cylindrical shells with random boundary and geometric imperfections, *Proceedings of ICOSSAR01*, June 2001, Swets & Zeitlinger, Lisse, CDrom

An approach to classify methods to control uncertainty in load-carrying structures

Roland Engelhardt[1,a], Jan Koenen[1,b], Matthias Brenneis[2,c], Hermann Kloberdanz[1,d], Andrea Bohn[1,e]

[1] Technische Universität Darmstadt, Product Development and Machine Elements PMD, Magdalenenstrasse 4, 64289 Darmstadt, Germany

[2] Technische Universität Darmstadt, System Reliability and Machine Acoustics SzM, Magdalenenstrasse 4, 64289 Darmstadt, Germany

[3] Technische Universität Darmstadt, Production Engineering and Forming Machines PtU, Petersenstrasse 30, 64293 Darmstadt, Germany

[a]engelhardt@pmd.tu-darmstadt.de, [b]koenen@szm.tu-darmstadt.de, [c]brenneis@ptu.tu-darmstadt.de, [d]kloberdanz@pmd.tu-darmstadt.de, [e]bohn@pmd.tu-darmstadt.de

Keywords: Methods, Classification, Uncertainty, Model of Uncertainty, Control of Uncertainty

Abstract Today, a wide variety of methods to deal with uncertainty in load-carrying system exists. Thereby, uncertainty may result from not or only partially determined process properties. The present article proposes a classification of methods to control uncertainty in load-carrying systems from different disciplines within mechanical engineering. Therefore, several methods were collected, analysed and systematically classified concerning their characteristic into the proposed classification. First, the classification differs between degrees of uncertainty according to the model of uncertainty developed in the Collaborative Research Centre CRC 805. Second, the classification differs between the aim of the respective method to descriptive methods, evaluative methods or methods to design a system considering uncertainty. The classification should allow choosing appropriate methods during product and process development and thus to control uncertainty in a systematic and holistic approach.

Introduction

Considering the life cycles of mechanical load-carrying products, it becomes obvious that both the product properties and their use differ in many ways from the assumed values. All variations resulting from more or less imprecisely determinable processes can be characterised and described as an uncertainty. During its life cycle, every mechanical load-carrying product such as truss structures, hydraulic cylinders, columns etc. is exposed to a multitude of different uncertainties. Material defects, production faults and unexpectedly high loads are typical for the field of engineering [1]. According to the definition, uncertainty in technical systems exists when product and process properties are undetermined and deviations from these properties occur [2]. Each uncertainty has to be taken into account, since having an impact on possibilities of utilisation and potentially being decisive for the safety of customers as well as for the environment. The activities were carried out commonly at the Collaborative Research Centre (CRC) 805: „Control of Uncertainty in Load-Carrying Structures in Mechanical Engineering". The aims are a significant increase of safety, reliability and efficiency in development, production and use as well as to save resources [3]. At the research centre uncertainty is regarded as described in the working hypothesis: *"Uncertainty occurs when process properties of a system cannot or only partially be determined. Uncertainty can be described and can be quantified with known methods of risk analysis"* [3]. Research at CRC 805 finally has the aim to control uncertainty in load-carrying structures. Their general problem is that uncertainty in loads and strength can have a significant impact. In product development, production and use many methods which deal with individual aspects of uncertainty are known. Especially for load-carrying structures, a coherent methodology of harmonised methods and proceedings is missing. This knowledge has to be transferred to load-carrying structures. The following demonstrates that the classification of methods makes its contribution in order to detect uncertainty and its cause systematically, to analyse it and finally to find robust solutions.

Overview of models and methods to deal with uncertainty

Uncertainty generally occurs in processes during the product life cycle of load-carrying structures. Especially loads and strength of the structure plays an important role for lifetime, reliability etc. To control uncertainty, design engineers of load-carrying structures always try to prove functional suitability during product development by means of computational or experimental examination of the operational stability and reliability. According to VDI 4001 reliability describes the probability of an item to perform its required function without failure under stated conditions for a stated period of time [4]. Thus, to predict reliability of load-carrying structures, it is essential to know real loads and strength. However, especially in the early stages of product development knowledge about conditions of use or expected lifetime of a future product is still very incomplete and a high degree of uncertainty regarding the expected processes and the properties of the product exists [5]. For example, material properties may be well known, but no geometries or even permissible deviations due to manufacturing processes are defined. Often, operational loads are only estimated by standards or simple calculations, but no load spectra are known [6]. In real products and real manufacturing processes, uncertainty exists in form of variance in product properties such as bearable loads or changing exploitation processes of products.

The exploration and classification of uncertainty is mainly considered in the field of decision theory in business studies. Due to increasing levels of information, uncertainty can be present in different degrees depending on the availability of information. The decision theory provides a possibility to deal with uncertainty by describing and evaluating action alternatives to find an optimal solution for a specific problem [7]. Nevertheless, this mainly qualitative approach is not sufficient to control uncertainty in load-carrying structures in a holistic sense. Therefore, highly diverse qualitative and quantitative methods can be used.

Classifications of uncertainty in uncertainty models are carried out in different disciplines. They include various uncertainties as well. As an uncertainty model is used for mapping different degrees of uncertainty to several classifications, it should be workable and applicable to the entire product life cycle of load-carrying structures. First research activities on the field of uncertainty in load-carrying structures proved that uncertainty can differ in a way that a differentiated consideration is necessary. For this purpose a model of uncertainty in terms of a classification of uncertainty was developed. The uncertainty model developed and used in the CRC 805 fulfils this requirement. Its categories will be described in the following.

Model of Uncertainty: in CRC 805 uncertainty is divided into three categories of uncertainty according to the increasing state of information about the probability distribution of the value of an uncertain product or process property: unknown uncertainty, estimated uncertainty and stochastic uncertainty see Fig. 1 [2]. The enumeration of this classification is elected to the degree of increasing information.

Unknown Uncertainty describes the situation of unknown deviations of a regarded property of an uncertain process. Based on this state of information, no comprehensive decisions can be made concerning the control of uncertainty. Unknown uncertainty often occurs in the beginning of product development when only little information about a future product is known and product properties are not determined yet [2].

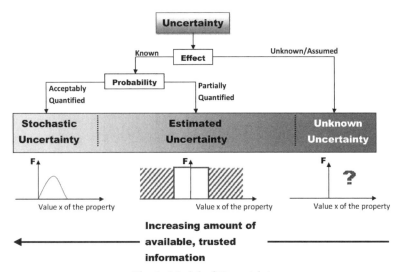

Fig. 1: Model of Uncertainty

Estimated Uncertainty describes a situation in which the effects of a regarded uncertain property are known. However, the probability distribution of the resulting deviation is only known partially. This is for example the case when incomplete information about expected properties of a product is known during product development or if, during manufacturing, product properties are analysed randomly only [2].

Stochastic uncertainty occurs when effects and resulting deviations of a regarded uncertain property are sufficiently (ideally completely) described by a probability distribution. Stochastic uncertainty is present after extensive analyses of properties in terms of quantifiable experiments and measurements [2].

When differentiating between these three categories, no sharp boundary can be drawn. The transition between the categories is fluent. As a general rule, uncertainty moves towards stochastic uncertainty if the amount of available and secure information increases.

Methods to deal with uncertainty: During development, production and use of load-carrying structures methods are used e.g. in order to reduce a lack of information, to evaluate products by risk or to optimise their strength against widely ranged loads. These qualitative and quantitative methods can be used for load-carrying structures, but in general also for other products.
A large variety of methods was collected. Out of these however, only few are mentioned representatively in this report. In order to control uncertainty in load-carrying structures, methods from the field of product development are used. Exemplarily listed: brainstorming, Failure Mode and Effect Analysis (FMEA), geometrical visualisations, active systems, sensitivity analyses, robust optimisation, ontology's, knowledge databases, design principles or solution catalogues [8,9,10].
In the field of production, methods like process planning, statistical design of experiment, simulations of cutting force, model of rigidity or optimisation of tools are applied [11,12,13].
Whereas in the field of the usage methods like Monitoring, Fuzzy-Logic, Shape Control, robust control, reliability-based design optimisation or methods from structural mechanics are applied [14,15,16].
To control uncertainty finally, robust product- and process-solutions have to be developed. In order to control uncertainty, applying only one method out of these fields is insufficient. Unknown uncertainty for example can be detected by the Delphi Method (consultations with experts) but not

assessed according to relevance. Therefore evaluation methods like the Failure-Mode- and Effect-Analysis are necessary. Evaluating influences of errors according to risk becomes possible [17]. The FMEA does not give any statements about how to make the system robust to these influences [17]. This can be achieved by solution catalogues or robust mathematical optimisation, which are however merely of a supporting nature.

Currently there are appropriate methods for single tasks and fields which can separately contribute to control uncertainty in load-carrying structures. The problem lies in the lack of a strategy that indicates when to use which method reasonably and how to link the methods.

In the field of product development, the development methodology VDI guideline 2221 provides a strategy for the systematic use of methods in order to develop new products [18].

The VDI 2221 proposes a general, industry independent proceeding model for development of technical products. The procedure consists of seven steps with its respective using methods, which are assigned to the same number of results. The guideline consists of a variety of methods which are described in order to obtain a structured approach [7,18].

Classifications of Methods

All methods mentioned in this chapter deal with uncertainty in different usually very specific ways. Some methods only detect and characterise uncertainty. Evaluation methods operate with correlations in products and processes and they offer a prediction about uncertainty with a high relevance. But uncertainty must be detected and characterised in advance in order to apply these methods. Finally there are methods, which offer system solutions for avoiding influences of uncertainty on the product or its production. Such solutions are not always satisfying because in most cases they restrict the usage possibilities.

To control uncertainty and find robust solutions a number of methods have to be systematically selected and used in combination. That means methods have to be combined in a holistic procedure. The mentioned methods have to be combined in order to consider and control uncertainty rather in a holistic way than partially. A systematic categorisation of these methods is necessary in order to get an overview of existing methods and an appropriate possibility to combine them. As described in the chapter "Model of Uncertainty", in the fields of uncertainty research, such a methodology does not exist.

For the classification of methods to control uncertainty two fundamental aspects have to be taken into account:
- Depending on the level of information according to the uncertainty model, uncertainty can occur very differently.
- Controlling uncertainty is a very complex problem that has to be solved in a structured way.

Thus, a successful classification of methods to control uncertainty requires a classification to two categories: Degree of uncertainty and phase in the process to control uncertainty.

A classification of uncertainty methods fulfilling both aspects is shown below.

Classification of methods according to the degree of uncertainty: For example the variability of working load at a new development can be completely unknown. The existent uncertainty can be classified as unknown uncertainty. In this case for instance, the method of Brainstorming can be used to increase the degree of information. Of little use would be the Monte Carlo simulation. It needs very exact input data to calculate meaningfully.

On the other hand, in the range of stochastic uncertainty, if variation is already existent in distribution functions, the Monte Carlo simulation can be used to determine influences of uncertainty. It is obvious that the uncertainty model can be used to classify methods. Thus it is possible to easily select the methods for a useful/successful application. But the classification according to the uncertainty model is not sufficient.

This general structuring of the approach to control uncertainty is based on the VDI 2221 [19]. A successful classification of methods to control uncertainty thus requires a classification to two categories. In this paper these categories are presented as horizontal degree of uncertainty and vertical phases.

In the horizontal degree of uncertainty, methods are divided into the three categories unknown uncertainty, estimated uncertainty and stochastic uncertainty according to the model of uncertainty (see chapter "Model of Uncertainty") [2]. Depending on the level of knowledge concerning information about uncertainty, different methods are used. Methods for the field of unknown uncertainty can be e.g. Brainstorming or expert talks. If a detailed acknowledgement is given and the uncertainty is located in an area of statistical uncertainty, the FEM-Calculation or monitoring-methods can be used.

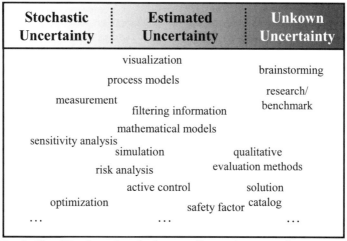

Fig. 2: Classification of methods according to the uncertainty categories

Classification of Methods in several phases of control of uncertainty: To solve the complex problem of controlling uncertainty several phases have to be proceeded:
- A full identification and description of uncertainty is necessary, which, if required, includes the whole product life cycle and the configuration of the involved systems.
- Since in general the number of examined uncertainty-afflicted properties is very high an evaluation of the uncertainty has to be done to enable a concentration on the relevant uncertainty.
- Before having performed the first two steps it is impossible to define a system which is uncertainty compatible.

In order to comprehensively control uncertainty the methods cannot be classified according to their data base, but to their classification in the general approach to control uncertainty. For this classification the vertical axis had been chosen.

The classification according to the different phases of uncertainty control, here, is illustrated in vertical orientation. This classification had been improved with conclusions gained out of the cooperation of CRC 805 and its subprojects in the research fields of production and utilisation. So the VDI guideline starts with a description of the system. Also in the collection of methods there are methods, which only have a descriptive character. This general structuring of the approach to control uncertainty is based on the VDI 2221 [19]

Methods to detect and describe uncertainty (Fig. 3): These are methods to identify and describe uncertainty. Also to provide information which analyse uncertainty or describe models. These models are used as a basis for the following method of evaluation.

		Stochastic Uncertainty	Estimated Uncertainty	Unkown Uncertainty
Methods to detect and describe uncertainty	Goal+target state		visualization process models	brainstorming
	Environment Influencing variable	measurement	filtering information	research/ benchmark
	Model connections		mathematical models	

Fig. 3: Methods to describe uncertainty

The description of methods can be subdivided. Orientated on the model of technical processes during the description of uncertainty three principle cases can be distinguished:
- Uncertainty as variations in the final condition of the processes, which represents the intended performance of the system.
- Uncertainty which occurs due to the influences of the environment to the system.
- Uncertainty in consequence of inner system correlation.

So methods to describe the Environment and target state fall in the category of descriptive methods. These are methods to characterise the process target and the process environment. Anymore the target state of the system should be included (e.g. list of requirements, property catalogue). Furthermore methods exist to characterise the **influencing variables** (e.g. disturbance variable, information, human) and methods for **Model correlations.** They describe inner process parameters and mathematical correlations in the system.

Methods to evaluate uncertainty (Fig 4): The second phase of the approach to control uncertainty includes methods to evaluate and estimate uncertainty. This deals with, for example, finding methods which can prioritize uncertainty and determine whether the effect of uncertainty is critical or not. These methods can clarify effects of arising uncertainty in processes. These can be, for example, simulation methods (which evaluate the system and its impacts) or risk evaluating methods (which include the environment). They are used as a basis for problem solving, which can find an optimal system for uncertainty.

		Stochastic Uncertainty	Estimated Uncertainty	Unkown Uncertainty
Methods to evaluate uncertainty	Product and process estimation	sensitivity analysis	simulation	qualitative evaluation methods
	Environment interaction		risk analysis	

Fig. 4: Methods to evaluate uncertainty

These methods can be divided into two categories. In category one, the **product and process** correlations of a load carrying system are estimated. This estimation neglects external circumstances and consequences (e.g. simulation methods).

Across the examination of the internal behaviour of the system, the environment interactions of the system can be used to evaluate the uncertainty. Examples for this are: not sufficiently known internal system behaviour during early stages or the evaluation of the use of the system. The second category contains methods to involve the **environment interaction** of the product. These could be methods such as risk assessment or reliability analysis. Target values used for evaluation and evaluation criteria are defined and thus scored upon. An individual assessment is done to check whether the effects of uncertainty are critical or not.

Methods used for system design under uncertainty (Fig. 5): These methods are used to react to uncertainty. Arrangements are defined to evaluate how resistant a system can be against uncertainty influences e.g. with robust processes. Thereby the focus lies on the avoidance or elimination of uncertainty in processes or on the adaptation of the system to existing uncertainty, so that the effects are no longer critical. These methods are, for example solution catalogues or methods of mathematical optimisation.

		Stochastic Uncertainty	Estimated Uncertainty	Unkown Uncertainty
Methods for system design under uncertainty	Avoidance/ elimination	safety factor		
	Adaption	optimization	active control	solution catalog

Fig. 5: Methods used for system design under uncertainty

If methods ought to be more robust, there are two general approaches. One approach **methods used to eliminate uncertain process properties**. Thereby information found by the characterisation and evaluation is used to make the process more robust. Weaknesses which are identified through evaluation are being eliminated. Normally strategies for avoiding and eliminating uncertainties in a process are characterised in VDI2221 [19]. The other option **Methods used for adaption to uncertain process properties**. The information which is found in the course of the characterisation and evaluation is applied. Normally strategies for structuring the process are characterised to make it robust towards uncertainty. Based on input parameters the process is calculated and optimised.

The combination of both described classifications results in a field in which the methods of uncertainty control can be reasonable sorted into. The described classification helps to put already known methods into a meaningful relationship. Finally methods from all three domains (description, evaluation and system design) are used to control uncertainty.

Utilisation of the method classification during the development process: On the other hand the classification extensively characterise all situations which occur during the handling of uncertainty. Depending on the level of information the degree of uncertainty can be estimated and depending on the progress of the procedure to control uncertainty the actual phase can be defined. Then, with the help of the method classification, an adequate method can easily be identified.

			Stochastic Uncertainty	Estimated Uncertainty	Unkown Uncertainty
Methods to control Uncertainty	Methods to describe uncertainty	Goal + target state		visualisation process models	brainstorming
		Environment Influencing variable		measurement filtering information	research/ benchmark
		Model connections		mathematical models	
	Methods to evaluate uncertainty	System behaviour		simulation sensitivity analysis qualitative evaluation methods	
		Environment interaction		risk analysis	
	Methods for system design under uncertainty	Avoidance/ elimination			safety factor
		Adaption		active control optimization production families	

Fig. 6: Summary of some methods to control uncertainty

First analyses of developed products have shown, that basic strategies or typical procedures for the orientation during the methodical control of uncertainty can be recommended:
- The control of uncertainty generally begins with the identification and description of uncertainty. Afterwards the evaluation of uncertainty follows, before methods to design a system uncertainty can be used suitably.
- Depending on the task (for example complexity of the product and units to produce) it can be decided on which level of uncertainty solutions should be developed. (At very high numbers of units, e.g. the level of uncertainty should be reduced to stochastic uncertainty). Principally procedures have been proven themselves which increase the information at the beginning of the development with the help of methods to identify and describe uncertainty so far. That work can be continued with an acceptable level of uncertainty.
- Only if a reasonable level of uncertainty is reached the application of matching methods for evaluating and designing the system should follow.

For a complete control of uncertainty it is necessary not just to move along the vertical axis in direction of system design, but to try to use methods to generate more information (horizontal axis) and to move in the direction of stochastic uncertainty. This has developed two basic strategies.

    1. One could try to get more information with the help of methods, to get a more detailed data basis. This could be done for example with monitoring methods to determine the loads with low degree of uncertainty which appear in the load-bearing system during the handling.

    2. It can be reasonable to stick to the existing data basis and (referring to this data basis) to make the process so robust, that uncertainty has few consequences. This can be done e.g. with an adaptronic control, which, during unknown loads, divides the loads to other components.

A guide to select the appropriate strategy is not part of this paper and is still under investigation.

Example: Path through the proposed classification of methods in the development process of a load-carrying system

To support the plausibility of the classification of methods, an example will be presented. This chapter discusses the development process of a simple load-carrying structure, in this case a bicycle seat post, with its different methods in terms of uncertainty concerning its safety and functionality. The connection of several methods used during the development process and their classification as proposed above are shown in Fig. 7 and are discussed below.

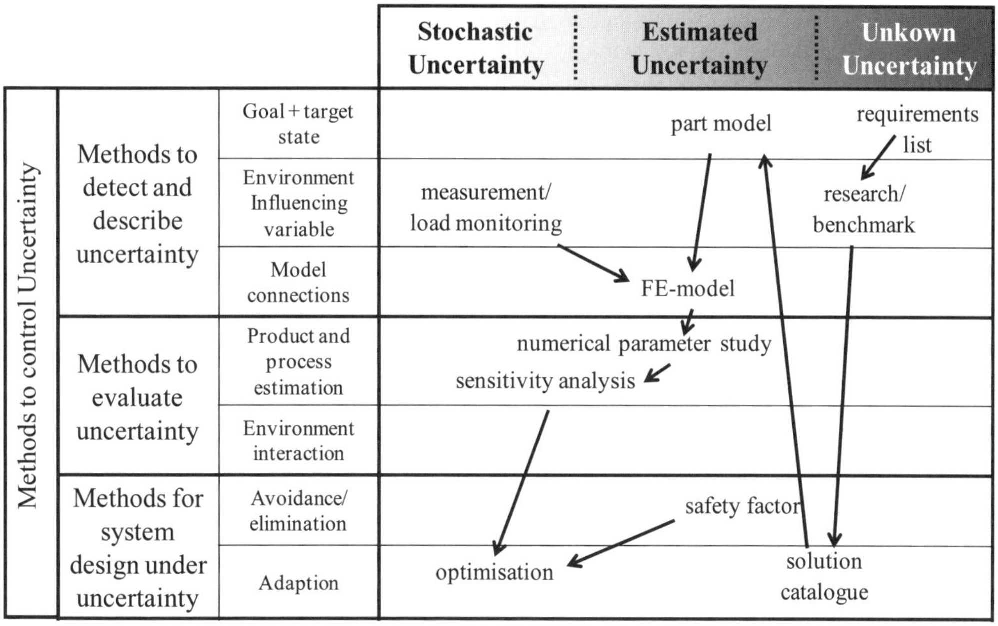

			Stochastic Uncertainty	Estimated Uncertainty	Unkown Uncertainty
Methods to control Uncertainty	Methods to detect and describe uncertainty	Goal + target state		part model	requirements list
		Environment Influencing variable	measurement/ load monitoring		research/ benchmark
		Model connections		FE-model	
	Methods to evaluate uncertainty	Product and process estimation	sensitivity analysis	numerical parameter study	
		Environment interaction			
	Methods for system design under uncertainty	Avoidance/ elimination		safety factor	
		Adaption	optimisation		solution catalogue

Fig. 7: Example for a connection of different methods in development process and its classification

At the beginning of a development process unknown uncertainty is present. For example, no operational loads are known, the geometry or even the material of the seat post is not specified yet. To clarify its specification, a requirement list provides information of the product to be developed, e.g. estimations on operational demands or on manufacturing costs and sales price, see Fig. 8a). Some research or benchmark on other seat posts on the market as well as simple calculations (for example multiplying the expected weight of the cyclist with a dynamical load factor), can help to gain a better understanding of environmental influences and thus help to identify occurring uncertainty for the development of the product. Both methods provide an increase of information about the demands on the seat-post. Based on this mainly qualitative information, a concept for the system design can be found (for example by the use of solution catalogue). For example, the principal design of the seat post and the material can be selected. Thereby, the degree of uncertainty is still high, but a first important step to control uncertainty is done due to the choice of a suitable concept, e.g. using methods of robust design [20], see Fig. 8b).

However, the next step is to visualize the seat post by detailing the concept in a part model, e.g. a CAD-model. Thus, a determined geometry may be defined. A finite-element-model provides a mathematical model to describe the relation between outer loads and inner stress or strain. The better loads are known, the better the structure can be designed to its operational demands. Determining loads by measurement (e.g. due to a load-monitoring in field trials) allows obtaining a significant increase of information on the loads with a low degree of uncertainty (e.g. due to a fully known frequency distribution of the loads), see Fig. 8c). The finite-element-model and loads are used in a numerical parameter study to evaluate the stress and strain within the seat post, see Fig. 8d). Sensitivity analysis provides a possibility to evaluate the effect of variance in product properties with respect to stress or deformation of the seat post, which is similar to an increase of available information. Those variances may occur for example due to scattering in geometrical and material properties resulting from fluctuations in the manufacturing process or scattering in loads and load directions, see Fig. 8e). The sensitivities can be used to optimize the geometry of the structure, for instance on one hand to minimize stress or deformation. On the other hand it is possible to reduce the sensitivity of stress or deformation with respect to uncertain values (e.g. the load direction) by finding a robust optimal design and thus to make a structure robust against uncertainty. However, some unavoidable uncertainty will remain, which can be eliminated by using appropriate safety factors (for example to prevent a failure due to a heavy weight cyclist).

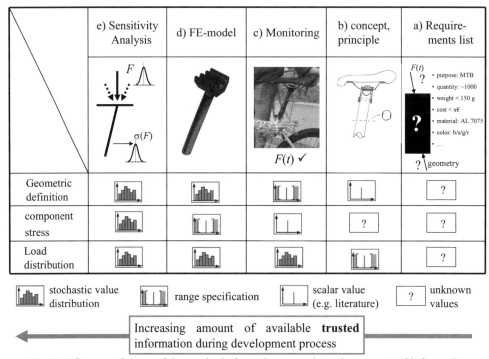

Fig. 8: Influence of some of the methods from the example to the amount of information

In summary, it is striking that in the context of the development process of a load-carrying structure, the path from unknown uncertainty is carried out over estimated uncertainty to stochastic uncertainty by sequential processing descriptive and evaluative methods to increase the amount of available trusted information and thus to control uncertainty. Finally, a concretion by methods for system design is achieved. This elementary example shows how uncertainty in the development process can be handled through the use of well-known methods. It is also illustrated, how uncertainty is controlled by the increasing amount of trusted information on the properties like geometrical parameters or operational loads of the load-carrying structure.

5 Conclusions and Outlook

The paper focused a classification of methods with the aim to control uncertainty in load-carrying structure within the entire product life cycle, thus product development, production, usage and recycling. The objective was to structure different known and even incumbent methods to deal with uncertainty into a plausible classification. It was shown, that methods to control uncertainty can be divided at the one hand concerning the three degrees of uncertainty (unknown uncertainty, estimated uncertainty and stochastic uncertainty). At the other hand, methods to control uncertainty can be divided into methods to describe uncertainty, methods to evaluate uncertainty and methods to design a load-carrying structure considering uncertainty. The classification can be used for orientation during the methodical approach and the selection of suitable methods. To support the plausibility of the classification, a development process of a bicycle seat was pointed out. Thereby, several methods were proposed to carry out unknown uncertainty into stochastic uncertainty by sequential processing descriptive and evaluative methods to control uncertainty.

Acknowledgement

We like to thank the Deutsche Forschungsgemeinschaft (DFG) for funding this project within the Collaborative Research Centre (SFB) 805.

References

[1] Bertsche, B.; Lechner, G.: *Zuverlässigkeit im Fahrzeug- und Maschinenbau – Reliability in Vehicle and Mechanical Engineering*. Berlin, Heidelberg, New York: Springer, 2004.

[2] Engelhardt, R., Koenen, J., Enss, G., Sichau, A., Platz, R., Kloberdanz, H., Birkhofer, H. and Hanselka, H.: *A Model to Categorise Uncertainty in Load-Carrying Systems*. In: Proceedings of MMEP Conference, Cambridge, July 2010.

[3] Hanselka, H., Platz, R.: *Ansätze und Maßnahmen zur Beherrschung von Unsicherheit in lasttragenden Systemen des Maschinenbaus – Controlling Uncertainties in Load Carrying Systems*, Konstruktion, 6, pp. 55–62, Springer VDI-Verlag, (2010).

[4] VDI 4001 Standard, *General guide to the VDI-handbook reliability engineering* VDI-standard, Beuth Verlag, Berlin, Wien, Zürich, (1985)-10

[5] Birkhofer, H.: *There Is Nothing As Practical As A Good Theory – An Attempt To Deal With The Gap Between Design Research And Design Practice*. In: Proceedings of the International Design Conference, May 2004.

[6] Haibach, Erwin: *Betriebsfestigkeit: Verfahren und Daten zur Bauteilberechnung - Structural durability: Methods and data for components calculation*, 3rd edition, Springer, (2006).

[7] Pahl, G., Beitz, W., Feldhusen, J., Grote, K.: *Engineering Design - A Systematic Approach*. Springer-Verlag, London, 2007.

[8] Saltelli et al.: *Global Sensitivity Analysis – a primer*. Chichester: John Wiley & Sons Ltd, 2008.

[10] Metropolis, N.; Ulam S.: *The Monte Carlo Method*. Journal of the American Statistical Association; Vol. 44, No. 247 (Sep., 1949), pp. 335-341

[11] Siebertz, K.; van Bebber, D.; Hochkirchen, T.: *Statistische Versuchsplanung – Design of Experiments (DOE)*; Springer-Verlag, Berlin Heidelberg, 2010

[12] Zabel, A.: *Prozesssimulation in der Zerspanung – Modellierung von Dreh- und Fräsprozessen – Process Simulation in machining – Modeling Turning and Milling Processes*, Habilitationsschrift TU Dortmund, Vulkan Verlag, Essen, 2010

[13] Schulz, H.; Emrich, A.K.: *Using the Principle of Genetic Algorithm (GA) für the Optimization of the Chip Flute of Drilling Tools*, Production Engineering VIII (2), 2001

[14] Adams, D. E.: *Health Monitoring of Structural Materials and Components*, John Wiley & Sons Ltd., 2007

[15] T. Haag, J. Herrmann, M. Hanss: *An identification procedure for epistemic uncertainties using inverse fuzzy arithmetic*, In: Proceedings of ISMA Conference 2010, pp. 5261ff.

[16] Kokkolaras, Mourelatos, Papalambros, *Impact of uncertainty quantification on design: an engine optimisation case study*, Int. journal of reliability and safety, Vol. 1, Nos. 1 / 2, 2006

[17] Grantham Lough, K., Stone, R., Tumer, I.Y.: *The risk in early design method.* In: Journal of Engineering design, 20(2009)2, pp. 155-173.

[18] VDI 2221: *Methodik zum Entwickeln und Konstruieren technischer Systeme und Produkte – Methodology to Develop and Design Technical Systems and Products.* Beuth-Verlag, Düsseldorf,1993.

[19] Mathias, J, Kloberdanz, H, Engelhardt, R., Birkhofer, H.: *Integrated Product and Process Development based on Robust Design Methodology.* In: ICED-International Conference on Engineering Design, 24.08.2009, Stanford, USA.

Time-dependent fuzzy stochastic reliability analysis of structures

Wolfgang Graf[1,a], Jan-Uwe Sickert[1,b]

Institute for Structural Analysis, Technische Universität Dresden, Germany

[a]wolfgang.graf@tu-dresden.de, [b]jan-uwe.sickert@tu-dresden.de

Keywords: Uncertainty, Imprecision, Imprecise Probability, Fuzzy Randomness, Reliability Analysis

Abstract. The paper reviews the development of reliability assessment in structural analysis under consideration of the non-traditional uncertainty model fuzzy randomness. Starting from a discussion of sources of variability and imprecision, uncertainty models are introduced. On this basis, numerical approaches are displayed for uncertain structural analysis and reliability assessment. Thereby, variations in time are considered which results in a time-dependent reliability measure. Capacity and applicability of the approaches are demonstrated by means of an example.

Introduction

The responses of structural systems which are subjected by external demands is realistically predictable only in an uncertain manner. This follows on one hand from the uncertainty in the external demands, e.g. loads, temperature, humidity. On the other hand, load bearing behavior of structural systems is uncertain also in the theoretical case of certain demands. As examples, uncertainty is observable in stiffness, strength, load redistribution capabilities and long-term phenomena. Different sources are accountable for this uncertainty, e.g. inhomogeneities, information deficit and systematical errors.

Traditional, uncertainty is considered in structural analysis and reliability assessment by means of probability theory without regard to the sources of uncertainty. In result, probabilistic measures are applied for the evaluation of structural systems. However, these reliability measures suggest more information as really included because of the frequentistic interpretation of probability which is established. That means, in the mind of most peoples the probability is connected with mass phenomenons like a game of dice. In engineering, however, we have to deal with structural systems which are produced with varying boundary conditions. Furthermore, unique specimens are regarded frequently. This is the reason that statistical validation of probability models must fail.

In addition, the observation of experiments is connected with uncertainty caused by an information deficit. Sources can be measurement devices with limited accuracy, e.g. digital devices, or measurements under dubious and unknown conditions. An example for the latter is the thickness of a structural member with rough surfaces. Further, subjective assessment is included into the evaluation of structural systems. Both, information deficit and subjectivity lead to imprecision. A probabilistic quantification of this imprecision would introduce unwarranted information in form of the probability density function. Even if an uniform distribution is applied, a certain regularity is presumed which represents more information then is available in reality. Therefore, imprecision should be modelled adequately by non-traditional uncertainty models. Different uncertainty models are suggested in the paper.

With the introduction of non-traditional uncertainty models including quantification methods, the questions arise concerning the consideration of this models within the structural analysis and reliability assessment. The paper provides answers to this questions in terms of applicable approaches for reliability assessment of structural systems. Special attention is given the consideration of time-dependent behavior. An example is added, in order to demonstrate the algorithms and its applicability.

Uncertainty models

According to their source, uncertainty can be subdivided in epistemic and aleatory uncertainty, see e.g. [1], [2].

Epistemic uncertainty results from incomplete knowledge concerning the real value of a structural parameter at a certain point in space and time, if it is assumed that a deterministic value exists. In other words an information gap occurs which could be reduced by generating additional background. Synonymous words for epistemic uncertainty are subjective, reducible and type-B uncertainty.

The aleatory uncertainty is the result of system inherent variations in space and time. That means the behavior of a system/structure varies from point to point and from time to time due to heterogeneity. It is also referred to as stochastic, not reducible and type-A uncertainty.

For adequate evaluation of structural reliability, aleatory and epistemic uncertainty have to be considered separately, in order to reflect both the information gap and the non reducible uncertainty in the results. This requirement motivates the discussion about different uncertainty models. Aleatory uncertainty can adequately described by stochastic models whereas for epistemic uncertainty non stochastic models are more appropriate. The latter is reasonable because of the subjectivity which impedes a stochastic description, see e.g. [3], [4].

Stochastic models presuppose the knowledge of underlying probability density function $f(x)$ and distribution function $F(x)$. If a sample with an sufficient large number of elements is available, the determination of $f(x)$ and $F(x)$ can be supported by descriptive statistics.

The epistemic uncertainty is also quantified by assessment function. Regarding its applicability, interval and fuzzy variables are suggested here. Intervals are valid, if only certain limits x_l and x_r are known and a further preference of values is impossible. The basis of interval analysis is an binary assignment of elements. Therefore, intervals are inappropriate, if a graduell assignment of elements to a set is possible, see e.g. [4]. With fuzzy sets, a grey scale is added to the black and white view of intervals. That means, elements can be assigned gradually to a fuzzy set according to

$$\tilde{A} = \{(x; \mu_A(x) | x \in \mathbb{X})\}. \tag{1}$$

Thereby, each element x of the fundamental set $\mathbb{X}$ is assessed by its membership $\mu_A(x)$. If $\mathbb{X}$ is a measurable variable, $\tilde{A}$ is also referred to as fuzzy variable $\tilde{x}$ assessed by a membership function $\mu_x(x)$ which could be convex or non convex. In most cases normalized functions $\mu_x(x)$ are applied with functional values in the interval $(0, 1]$. If $\mu_x(x)$ consists of two linear branches, the respective fuzzy variable is referred to as fuzzy triangular number and specified by the triple $< l;\ p;\ r >$ with the peak value p and the interval $[l,\ r]$ containing all x for which holds $\mu_x(x) > 0$.

For the numerical treatment the membership functions of fuzzy variables are discretized in a family of α-level sets. The α-levels define the minimum membership of all included elements.

$$X_{\alpha_k} = \{x \in \tilde{x} |\ \mu(x) \geq \alpha_k\} \tag{2}$$

Further details to fuzzy variables are contained in [5, 6, 7].

In many engineering applications aleatory and epistemic uncertainty arise at the same time. This was the motivation for the development of imprecise probability concepts, as for example described in [6, 8, 9, 10]. More basically are the works [11, 12, 13]. Joining both randomness and fuzziness in a numerical efficient manner, a fuzzy random variable $\tilde{X}$ can be defined on the basis of [6] as result of the mapping

$$\tilde{X} : \Omega \rightarrow \mathcal{F}(\mathbb{R}). \tag{3}$$

Thereby, $\mathcal{F}(\mathbb{R})$ denotes the set of all fuzzy quantities which exist in $\mathbb{R}$. A fuzzy random variable $\tilde{X}$, as defined in [6], can be expressed as a family of real-valued random variables X

$$\tilde{X} = \left(X_\varsigma^\alpha : \Omega \rightarrow \mathbb{R}\right)_{\varsigma \in I(\alpha)}^{\alpha \in (0,1]}, \tag{4}$$

where $I(\alpha)$ denotes an α-level set according to Eq. (2) and Ω the set of all elementary events representing possible certain results of a random experiment. The random variables X_ς^α are originals of $\tilde{X}$, see [6], and ς is a certain realization of a fuzzy bunch parameter $\tilde{\varsigma}$. Each original X_ς^α is specified by a probability distribution function $F_{X_\varsigma^\alpha}$ as known from the traditional probability theory. Therewith, the fuzzy probability distribution function of $\tilde{X}$ results in

$$F_{\tilde{X}} = \left(F_{X_\varsigma^\alpha}\right)_{\varsigma \in I(\alpha)}^{\alpha \in (0,1]} \quad . \tag{5}$$

Since, there is an infinite number of distribution functions $F_{X_\varsigma^\alpha}$, $F_{\tilde{X}}$ can be approximated by

$$F_{\tilde{X}} = \left(F_{X_{\underline{s}}^\alpha}\right)_{\underline{s} \in \underline{I}(\alpha)}^{\alpha \in (0,1]} \tag{6}$$

with the Cartesian product of α-level sets $\underline{I}(\alpha)$. Thereby, the arbitrary indicator ς is replaced by a more specific bunch parameter vector $\underline{s} \in \underline{\tilde{s}}$. This bunch parameter vector contains typical parameters of probability distribution functions which are specified as fuzzy variables, such as for example fuzzy mean and fuzzy standard deviation. As an more detailed example, the one-dimensional fuzzy probability distribution function of Gumbel type with two fuzzy bunch parameters $\tilde{s}_1$ and $\tilde{s}_2$ then reads

$$F_{\tilde{X}}(\underline{\tilde{s}}, \underline{t}) = exp\left(-exp(-\tilde{s}_1(x - \tilde{s}_2))\right). \tag{7}$$

Further examples for the specification of fuzzy probability distribution functions are published in [6, 7, 14].

The bunch parameter representation enables a decoupling of fuzziness and randomness, which provides the basis for the numerical realization of uncertain structural analysis and reliability assessment.

Uncertain data like, e.g., material parameters, loading or boundary conditions may be more generally described by fuzzy random functions. Based on the theories of fuzzy randomness, fuzzy sets, and probabilistic, a fuzzy random function $\tilde{X}(\underline{t})$ is defined as a family of fuzzy random variables $\tilde{X}_t$ according to Eq. (4) in the parameter space $\mathbb{T}$.

$$\tilde{X}(\underline{t}) = \{\tilde{X}_t = \tilde{X}(\underline{t}) \,\forall \underline{t} \,|\, \underline{t} \in \mathbb{T}\} \tag{8}$$

For numerical simulation in structural analysis a second definition of fuzzy random functions may be advantageously applied. In dependency of the fuzzy bunch parameters $\underline{\tilde{s}}$ a fuzzy random function then reads $\tilde{X}(\underline{t}) = X(\underline{\tilde{s}}, \underline{t})$ with

$$X(\underline{\tilde{s}}, \underline{t}) = \{(X_j(\underline{t}), \mu(X_j(\underline{t}))) \,|\, X_j(\underline{t}) = X(\underline{s}_j, \underline{t}); \mu(X_j(\underline{t})) = \mu(\underline{s}_j) \,\forall \underline{s}_j \in \underline{\tilde{s}}\} \tag{9}$$

A fuzzy random function is specified by their multi-dimensional fuzzy probability distribution function $F_{\tilde{X}}(\underline{x})$ and the accompanying multi-dimensional fuzzy probability density function $f_{\tilde{X}}(\underline{x})$.

In general, a fuzzy random function depends on time τ, spatial coordinates $\underline{\theta} = (\theta_1, \theta_2, \theta_3)$, and further parameters ϕ like temperature or humidity. The different coordinates are lumped together in the parameter vector $\underline{t} = (\underline{\theta}, \tau, \phi)$. A fuzzy random function is referred to as fuzzy random field if it solely depends on spatial coordinates $\underline{\theta}$. In the case of exclusive time-dependency, a fuzzy random process is created. Fuzzy random fields and processes are included as special case in the definition of fuzzy random functions.

Reliability assessment

From traditional time-independent safety concept with random variables, the introduction of limit states

$$g(\underline{x}) = R(\underline{x}) - S(\underline{x}) = 0 \tag{10}$$

is known. Thereby, $R(\underline{x})$ represents structural resistance and $S(\underline{x})$ expresses the loading. Furthermore, the failure probability P_f is computed by integration of the joint density function $f(x)$ in the failure domain $g(\underline{x}) \leq 0$ according to

$$P_f = \int_{\underline{x}|g(\underline{x})\leq 0} f_X(\underline{x})d\underline{x}. \tag{11}$$

Due to the application of the generalized uncertainty model fuzzy randomness, input variables are modeled as fuzzy random variables $\tilde{X}$. Now, the structural design aims at a sufficient reliability in terms of the fuzzy failure probability [6, 7, 15] for the event that holds $g(\underline{x}) \leq 0$. In the case that the imprecision of an input variable prevail over their randomness, it is modeled as fuzzy variable. This fuzzy variable affects the directly the limit state function. Then, the fuzzy limit state has to be evaluated whereby the fuzziness is represented by bunch parameters $\tilde{s}_g$. The fuzzy failure probability is computed with

$$\tilde{P}_f = \left\{ P_f = \int_{\underline{x}|g_T(\underline{s}_g,\underline{x})\leq 0} f_X(\underline{s},\underline{x})d\underline{x} \; \forall \; \underline{s} \in \tilde{\underline{s}}, \; \underline{s}_g \in \tilde{\underline{s}}_g \right\} \tag{12}$$

According to Eq. (12), $\tilde{P}_f$ represents an assessed set of real-valued failure probabilities specified by certain realizations of the fuzzy bunch parameter vectors $\tilde{\underline{s}}$ and $\tilde{\underline{s}}_g$.

In civil engineering practice, the safety index β, instead of the failure probability, is often used to quantify the reliability level. When all $X \in \tilde{X}$ has a normal distribution, $\tilde{\beta}$ is exactly determined by means of failure probability, given by

$$\tilde{\beta} = \left\{ \beta = \Phi^{-1}(1 - P_f) \; \forall \; P_f \in \tilde{P}_f \right\} \tag{13}$$

where Φ is the cumulative distribution function for the standard normal distribution. In the case where some $X \in \tilde{X}$ have other distributions, $\tilde{\beta}$ according to Eq. (13) is an approximation. However, the advantages are that larger values of $\tilde{\beta}$ corresponds to a higher reliability and the numerical effort of the computation of $\tilde{\beta}$ is smaller then for $\tilde{P}_f$, if the fuzzy first order reliability method [6, 16] is applied.

With the measures $\tilde{P}_f$ and $\tilde{\beta}$, the reliability may be considered in the design process of structural systems. On the one hand, the reliability can be maximized under consideration of e.g. economical constraints. On the other hand, the economical effort can be minimized without to violate reliability constraints. Therewith, the reliability can be a design objective or a design constraint.

The reliability measure $\tilde{P}_f$ is computed by a hierarchical algorithm, displayed in Fig. 1. The computational algorithm consists of fuzzy, stochastic and deterministic analysis covered by the quantification of uncertain variables and the evaluation of results.

The analysis algorithms are arranged in three nested loops as shown in Fig. 1. In the outer loop the membership function $\mu(\tilde{P}_f)$ of the fuzzy failure probability $\tilde{P}_f$ is computed, using the α-level optimization. Initially, all fuzzy bunch parameters $\tilde{\underline{s}}$ and all pure fuzzy quantities $\tilde{\underline{x}}$ are combined by the cartesian product. They form the space of all fuzzy parameters. On the basis of an evolutionary strategy points are selected for which the stochastic analysis (middle loop) has to be determined. During optimization, stochastic analysis is performed repeatedly. The membership function $\mu(P_f)$ is computed with the aid of the set of α-level bounds containing the minimum and the maximum values which belong to the respective α-level. More details could be found in [5].

In principle, all known algorithms for the computation of failure probability may be utilized within α-level optimization. However, in most practical cases simulation methods are appropriate. In [15] the Monte-Carlo simulation is chosen and extended to the fuzzy Monte-Carlo simulation (FMCS). In order to decrease the numerical effort of the FMCS so called variation reduction methods could be applied for the stochastic analysis. Examples are importance sampling [18] and subset sampling [19]. On this

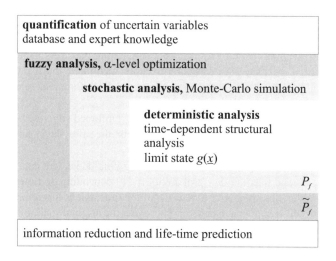

Fig. 1: Algorithm of fuzzy stochastic analysis for computation of fuzzy failure probability $\tilde{P}_f$

basis the fuzzy adaptive importance sampling method (FAIS) has been developed which is based on a modification of the adaptive importance sampling algorithm according to [18].

The deterministic structural analysis on the basis of a nonlinear FE model represents the core (inner loop) of the simulation. Thereby, the quality of the deterministic structural model determines the quality of $\tilde{P}_f(\tau_k)$ significantly. Results close to the reality can only be obtained, if the governing nonlinearities are considered within the deterministic time-dependent analysis.

The results of the latter serve as indicator

$$I(g(\underline{x})) = \begin{cases} 1 & \text{if } g(\underline{x}) \leq 0 \\ 0 & \text{if } g(\underline{x}) > 0 \end{cases} \tag{14}$$

whether a structure with the realized set of certain parameters fulfill the certain survival condition $g(\underline{x}) > 0$ or not. This information is applied in the Monte-Carlo simulation (MCS) to estimate a certain failure probability

$$\hat{P}_f = \frac{1}{N} \sum_{i=1}^{N} I(g(\underline{x})) \tag{15}$$

on the basis of N deterministic analysis. Eq. (15) only results in an estimator which tends to P_f for sufficient large N. The required number N can be determined by means of the Chebyshev's inequality in dependency of a predefined confidence level and the expected order of magnitude of P_f, se e.g. [17].

The computed fuzzy failure probability could be mapped onto a certain value by means of different algorithms of defuzzification, see [6]. However, valuable information is loss with this defuzzification. It is recommended by the authors, that engineers learn to deal with the fuzzy results and use it as basis for their decisions.

In order to consider fuzzy random field within the reliability analysis, the fuzzy stochastic finite element method (FSFEM) was introduced [7, 21]. Thereby, the Karhunen-Loeve expansion [22] is applied in the stochastic analysis.

Time-dependent reliability

The safety level of a structure is not constant over the lifetime, but time-dependent. For example alteration, damage, deterioration, fatigue or use for new purposes can lead to the reduction of the safety level. The situation is displayed in Fig. 2 with two fuzzy random processes $\tilde{R}(\tau)$ and $\tilde{S}(\tau)$. Both processes are functions with uncertain functional values, represented by a gray-scale chart for all time points τ. It is assumed that the requirements concerning the stress increases in time. On the other hand, the structural resistance is decreasing in time. At time τ_k the fuzzy probability density functions $\tilde{f}(R(\tau))$ and $\tilde{f}(S(\tau))$ of $\tilde{R}(\tau)$ and $\tilde{S}(\tau)$ are shown. Both functions represent an assessment of the associate uncertain functional values.

With strengthening (e.g. with application of thin textile reinforced concrete layers [15]) and maintenance the structural resistance can be increased. Therewith, the safety level can be raised again into the range of the accepted values.

An extended algorithm has been developed considering time-dependent data uncertainty of fuzzy randomness, see [15, 7]. The fuzzy failure probability $\tilde{P}_f(\tau)$ depending on time τ is now introduced as measure of the reliability. In order to compute $\tilde{P}_f(\tau)$ the time axis is discretized in points τ_k. Then $\tilde{P}_f(\tau_k)$ is computed for selected points in time τ_k. Therefore, the original space of the basic variables is constructed using all fuzzy random load, material, and geometry parameters which affect the reliability in τ_k. These parameters are mathematically specified by fuzzy random functions $\tilde{X}(\tau)$ or random functions $X(\tau)$. The discretization of these functions results in a set of correlated fuzzy random variables $\tilde{X}_t$ and random variables X_t, respectively. If we regard the X_t as special case of the $\tilde{X}_t$ then $\tilde{P}_f(\tau)$ is defined by

$$\tilde{P}_f(\tau = \tau_k) = \int_{\underline{x}|g_\tau(\tilde{\underline{s}}_g,\underline{x})\leq 0} f_t(\tilde{\underline{s}}, \underline{x})d\underline{x} \qquad (16)$$

with the fuzzy joint probability density function $f_t(\tilde{\underline{s}}, \underline{x})$ and the fuzzy limit state surface $g_\tau(\tilde{\underline{s}}_g, \underline{x}) = 0$ both in bunch parameter representation. For practical applications the integral including fuzzy bunch parameters is evaluated with simulation methods. Thereby the algorithm described in the previous section is repeatedly applied for all time points τ_k. More details for the fuzzy stochastic solution and examples have been presented in [7] and [20].

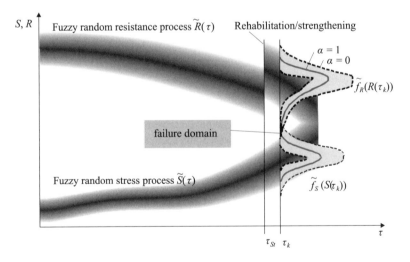

Fig. 2: Fuzzy random stress and resistance processes

The time-dependent approach could also be applied in the analysis of dynamic systems [23] and for the prediction of the lifetime of a structural system [24].

Example

As example the time-dependent chloride induced corrosion of reinforcement steel is investigated. The steel is embedded in concrete. If the critical chloride content in the environment of the steel is exceeded, corrosion may cause. Due to this corrosion, the thickness of the steel bars will be reduced. Regarding the load induced stress in the steel, the limit state of load bearing capacity can be determined under consideration of the residual cross section. If the limit state is reached, failure of the steel bars may occur.

In [25] a probabilistic model is shown which quantifies the corrosion of reinforcement steel in concrete caused by chloride ingress. The decrease of the reinforcement steel cross section is modeled by the decrease of diameter D depending on time τ

$$D(\tau) = D_O - 0.0232\,(\tau - T_i) \cdot i_{corr} \tag{17}$$

with the original diameter D_O, the corrosion rate i_{corr}, and the corrosion initiation time T_i. If C_{cr} is assumed to be the chloride corrosion threshold and x is the concrete cover thickness, then the corrosion initiation time T_i can be computed solving the differential equation

$$T_i = \frac{x^2}{4\,D_c} \cdot \left(\operatorname{erf}^{-1}\left(1 - \frac{C_{cr} - C_i}{C_O - C_i}\right) \right)^{-2} \tag{18}$$

of Ficks second law of diffusion. The diffusion coefficient D_c and the chloride surface content C_O in Eq. (18) are important variables in corrosion estimates. Different improved models were introduced in last decade in order to determine D_c as well as C_O. Here, a probabilistic model based on [25] is extended. Both D_c and C_O are modeled as fuzzy random variables $\tilde{D}_c$ and $\tilde{C}_O$. The chloride corrosion threshold C_{cr}, the initial chloride content C_i, and the concrete cover thickness x are assumed to be deterministic.

The mean values of the Gaussian distributed diffusion coefficient and chloride surface content are determined in dependence of the grade of deterioration G_D. As result of an inspection of a bridge experts evaluate the deterioration by the aid of a linguistic variable using the grades "low", "medium", or "high". However, the subjective assessment of different experts leads to different linguistic values G_D. Therefore, the mean values are modeled as fuzzy triangular numbers according to Fig. 3. Therewith,

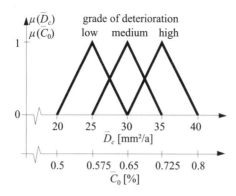

Fig. 3: Membership functions of the fuzzy mean values of $\tilde{D}_c$ and $\tilde{C}_O$ in dependency of the grade of deterioration

the subjectivity of the expert evaluation is quantified and will be mapped onto the fuzzy failure probability. The values with the membership $\mu = 1$ are conform to the suggestion given in [25]. The standard deviation is also taken from [25] as deterministic values $\sigma_{Dc} = 2.5$ mm/a and $\sigma_{C_0} = 0.038$. As a result of the functional dependency of $\tilde{T}_i$ on the fuzzy random variables $\tilde{D}_c$ and $\tilde{C}_O$ according to Eq. (18), also, the chloride initiation time becomes a fuzzy random variable $\tilde{T}_i$. Fig. 4 shows the probability distribution $\tilde{F}(T_i)$ of the fuzzy corrosion initiation time $\tilde{T}_i$. The solid line represents the certain result of a stochastic analysis without fuzziness. The gray area between the dashed lines captures all possible distribution functions which result due to the consideration of fuzzy mean values $\tilde{D}_c$ and $\tilde{C}_O$.

Additional to $\tilde{T}_i$ the corrosion rate i_{corr} is uncertain. [25] suggests to model i_{corr} as a Gaussian random variable and gives deterministic values for the coefficient of variation according to "typical values in normal environment". However, in most cases the evaluation of the environmental conditions by experts differs from one to another, i.e. the coefficient of variation is uncertain. Here, the corrosion rate is quantified by a Gaussian fuzzy random variable with the deterministic mean value $\tilde{i}_{corr} = 2$ A/cm and the fuzzy standard deviation modeled by the fuzzy triangular number $< 0.3; 0.4; 0.5 >$. The outcome of the consideration of the fuzzy random variables $\tilde{i}_{corr}$ und $\tilde{T}_i$ is the fuzzy random process $\tilde{D}(\tau)$ describing the time-dependent reinforcement steel diameter according to

$$\tilde{D}(\tau) = D_O - 0.0232\,(\tau - \tilde{T}_i) \cdot \tilde{i}_{corr} \tag{19}$$

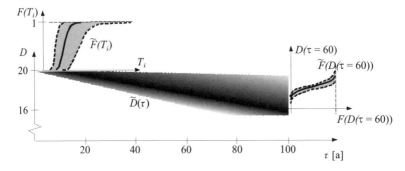

Fig. 4: Fuzzy random process $\tilde{D}(\tau)$ describing the time-dependent decrease of the diameter of a reinforcement steel with original diameter $D_O = 20$ mm

The fuzzy random process is shown in Fig. 4. The gray-scale chart represent then frequency of occurrence in time. On the right hand side, the fuzzy probability distribution function $\tilde{F}(D(\tau = 60a))$ is displayed as example. This function could be used to compute the fuzzy failure probability if a certain limit in terms of a required diameter is defined. Due to the consideration of fuzziness, not only a certain value is obtained but an assessed interval. This represent the benefit of the fuzzy stochastic reliability analysis, because the imprecision is visible in the results.

More examples concerning the time-dependent reliability analysis of also complex structures could be found the cited literature, e.g. in [7, 15, 24].

Summary

Coping with uncertainty in structural reliability analysis is only possible, if variability and imprecision are modelled separately. This ensures the reflection of the available information concerning the inputs also in the results of all computations especially in the reliability measure.

In order to perform reliability analysis, the uncertainty model fuzzy randomness is introduced. This model enables the jointed quantification of variability and imprecision. For numerical reason the bunch parameter representation has been introduced. On this basis, a numerical analysis algorithm is represented for the computation of fuzzy failure probability. Further, the paper contains an extension in order to consider time-dependent structural behavior.

The presented approach is numerically expensive, because of the repeated performance of Monte-Carlo simulation. Thereby, the computational time of the deterministic analysis is responsible for the total effort. In order to increase the efficiency of the algorithm, metamodels should be applied. Powerful metamodels could be created by means of artificial neural networks. An overview regarding the consideration of uncertainty in engineering is provided in [26] and [27].

Acknowledgment

The authors gratefully acknowledge the support of the German Research Foundation (DFG) in the framework of the Collaborative Research Center 528 ``Textile Reinforcement for Structural Strengthening and Repair".

References

[1] J. Helton, J. Johnson, W. Oberkampf and C. Storlied: Computational methods in applied mechanics and engineering 196 (2007), pp. 3980--3998

[2] A. Der Kiureghian: Probabilistic engineering mechanics, 23 (2008), pp. 351--358

[3] J. Helton and W. Oberkampf: Reliability Engineering & System Safety, 85 (2004), pp. 1--10

[4] B. Möller and M. Beer: Special Issue of Computers and Structures 86 (2008), pp. 1024--1041

[5] B. Möller, W. Graf and M. Beer: Computational Mechanics 26 (2000), pp. 547--565

[6] B. Möller and M. Beer *Fuzzy Randomness -- Uncertainty in Civil Engineering and Computational Mechanics* (Springer, Berlin 2004)

[7] J.-U. Sickert *Fuzzy-Zufallsfunktionen und ihre Anwendung bei der Tragwerksanalyse und Sicherheitsbeurteilung* (Veröff. Institut für Statik und Dynamik der Tragwerke, Heft 9, Techn. Univ. Dresden, 2005)

[8] R. Viertl *Statistical Methods for Fuzzy Data* (John Wiley & Sons, Chichester 2011)

[9] T. Augustin and R. Hable: Structural Safety 32 (2010), pp. 358--365

[10] P. Walley *Statistical reasoning with imprecise probabilities* (Chapman & Hall, London, 1991)

[11] H. Kwakernaak: Information Sciences 15 (1978), pp. 1--19

[12] H. Kwakernaak: Information Sciences 17 (1979), pp. 253--278

[13] M.L. Puri and D. Ralescu: Journal of Mathematical Analysis and Applications 114 (1986), pp.409--422

[14] A. Abdkader, W. Graf, B. Möller, P. Offermann and J.-U. Sickert: AUTEX Research Journal 2 (2002), pp. 115--125

[15] B. Möller, M. Beer, W. Graf and J.-U. Sickert: Computers & Structures 84 (2006), pp. 585--603

[16] B. Möller, W. Graf and M. Beer: Computers & Structures 81 (2003), pp. 1567--1582

[17] R. Rubinstein *Simulation and the Monte Carlo Method* (Wiley, New York, 1981)

[18] S. Mahadevan and P. Raghothamachar: Computers & Structures 77 (2000), pp. 725--734

[19] S.-K. Au and J. Beck: Probabilistic Engineering Mechanics 16 (2001),pp. 263--277

[20] S. Pannier, J.-U. Sickert, W. Graf and M. Kaliske, in: 2nd International Conference on Engineering Optimization (Lisbon 2010)

[21] J.-U. Sickert, B. Möller and W. Graf, in: Application of Statistics and Probability in Cicil Engineering -- Proceedings of the ICASP 10, edited by J. Kanda, T. Takada and H. Furuta (Taylor & Francis, 2007)

[22] C.A. Schenk and G.I. Schuëller *Uncertainty Assessment of Large Finite Element Systems* (Springer, Berlin Heidelberg, 2005)

[23] B. Möller, W. Graf, J.-U. Sickert and F. Steinigen: Mathematical and Computer Modelling of Dynamical Systems 15 (2009), pp. 515--534

[24] B. Möller, M. Liebscher, S. Pannier, W. Graf and J.-U. Sickert: Structure and Infrastructure Engineering: Maintenance, Management, Life-Cycle Design and Performance 7 (2011), pp. 325--340

[25] P. Thoft-Christensen, in: Proceedings of the Workshop on Structural Reliability in Bridge Engineering, (University of Colorado Boulder, 1996)

[26] Information on http://www.uncertainty-in-engineering.net

[27] W. Graf, S. Freitag, J.-U. Sickert, S. Pannier and M. Kaliske in: 2nd International Conference on Soft Computing Technology in Civil, Structural and Environmental Engineering (CSC-2), Chania, edited by Y. Tsompanakis and B.H.V. Topping (Saxe-Coburg Publications, Stirlingshire, 2011)

# Ontology-based Information Model for the Exchange of Uncertainty in Load Carrying Structures

André Sprenger [1, a], Michael Haydn [2, b], Serge Ondoua [3, c], Lucia Mosch [1, d], Reiner Anderl [1, e]

[1] Department of Computer Integrated Design (DiK), Technische Universität Darmstadt, Petersenstr. 30, 64287 Darmstadt, GERMANY

[2] Institute for Production Management, Technology and Machine Tools (PTW), Technische Universität Darmstadt, Petersenstr. 30, 64287 Darmstadt, GERMANY

[3] Department of System Reliability and Machine Acoustics (SzM), Technische Universität Darmstadt, Magdalenenstr. 4, 64289 Darmstadt, GERMANY

{[a]sprenger, [b]haydn, [c]ondoua, [d]mosch, [e]anderl}@sfb805.tu-darmstadt.de

Keywords: Uncertainty, Information Model, Collaboration

Abstract. Knowledge about future process properties is crucial for the development of safe and economic products with load carrying structures. Real processes are influenced by uncertainty what causes scattering and deviation from assumed values. As a consequence, products are often oversized or even product failures can occur. To control uncertainty, extensive knowledge about future processes is necessary in the development process. This paper shows an approach for the representation of uncertainty in production- and usage-processes, according to scattering properties and their cause and effect relations. This approach is used as a common platform for storing, locating, comparing and reuse of knowledge about uncertain properties and their relations. The core of the proposed approach is an ontology-based information model with the ability to represent different levels of trusted information in relation to process parameters and cause and effect relations.

Introduction

The development of load carrying structures is a knowledge intensive task. The dimensioning of such a product requires definite knowledge about product's stress and strength during future product operation. In product development the set values for the products properties are determined. The actual property values of the product are determined by the manufacturing and usage processes [1]. Thus, available knowledge about processes and its exchange between stakeholders is essential for a successful product development.

Nowadays, lots of information is collected with different methods like e.g. quality methods, process capability or usage monitoring. They all deal with uncertainty, but they have different specific descriptions of it. Furthermore these uncertain product and process properties are related to each other within different life cycle phases [1].

To represent this knowledge an ontology-based approach for an information model is introduced in this paper. This model allows the detailed representation of uncertain property values and the relationships between properties. In addition, an overview of semantic technology used in product development is given. Furthermore an information model for the representation and the exchange of uncertainty is presented. It is based on a process- and an uncertainty-model [2,3] and relates the uncertainty to the product model.

Two important examples clarify the implementation of the information model during a product development task. Both were conducted with a simple load carrying tripod system. The first example focuses on the tripod's manufacturing process chain, consisting of drilling and reaming. The second example considers the tripod's usage process with the analysis of load distribution in the three legs of the tripod. The tripod and the basic experiments are described in [2]. Knowledge resulting from these experiments is formalized for the representation in the information model. The ontology-based approach renders the possibility to represent relationships and effects between information items.

An important aspect for an engineer is the comprehensibility of the results of the ontology-system. This helps to gain the necessary benefits during the product development. Thus, a major requirement for the new ontology-based information model is an expedient and powerful visualization technique of the results. To give the engineer a common understanding of these results, new visualization techniques have to be developed and embedded into well-established Computer Aided (CA)-tools, especially CAD-systems (Computer Aided Design). The new designed visualization, integrated in a CAD-system, works on the basis of a three-dimensional parametric CAD-model. The concept of an uncertainty-browser, attached to CAD-systems, supports the clarity and comprehensibility of the related results.

State of the Art

In this chapter the basic models of uncertainty, the process model and their representation in an information model as well as basic semantic technology are presented. Furthermore two examples using the tripod system will be introduced.

Uncertainty model – a working hypothesis of the CRC 805

In the CRC 805 a model of uncertainty which provides three categories of uncertainty was developed. This model is presented in [3] and described in the following paragraph. Depending on the amount and quality of available information three types of uncertainty description can be distinguished: In the case of "unknown uncertainty" a process property and its interdependencies are not known or even ignored. Within this level of awareness, no distinct decision concerning a property can be made. With the type "estimated uncertainty" interdependencies could be identified but the quantification of the property is done by intervals, tolerances or nominal values. In case of "stochastic uncertainty" the lowest level of uncertainty is reached. Properties and their interdependencies are well-known and properties are described by frequency distributions. The classification of uncertainty in this model depends on the description of the uncertain process property and known effects to other process properties. There are distinct descriptions of uncertainty like distribution functions, different types of intervals and the interdependencies between process properties.

"Unknown and estimated uncertainty", as well as "stochastic uncertainty" were established in CRC 805 in consideration of the existing uncertainty definitions like e. g. epistemic uncertainty and aleatory uncertainty [3]. In comparison with those former existing uncertainty definitions, the uncertainty model of the CRC 805 aims to represent any uncertainty that may occur in the various processes during the product life cycle of a load carrying structure.

Exchange of uncertain process properties

A model, called "uncertainty data type" (UDT), for the representation of uncertain product and process properties in the field of mechanical engineering is introduced in [5]. Its aim is to represent the uncertainty of process/product properties in a digital model for virtual product creation purposes, focusing the exchangeability of those types of data. The approach is founded on the model of uncertainty of the CRC 805 and describes the named three types of uncertain properties.

"Stochastic uncertainty" is described by distribution functions or histograms. Distributions can be represented either by the parameters and key indicators of a common probability distribution function, like the normal distribution, or by "value pairs" if the function is unknown or from a not common type. The representation of histograms is also realized by these "value pairs" describing the class name, optional its width and the frequency. The "estimated uncertainty" is described by different types of intervals like "tolerance", "value interval" and a "nominal value" without tolerance information. For "unknown uncertainty" an empty value is given, showing that there is no information available. These representations of uncertainty are encoded in a common XML-scheme for a representation and exchange of uncertain process properties along the product life cycle.

Process model – a working hypothesis of the CRC 805

As described in the introduction, uncertainty occurs in processes within all phases of the product life cycle and influences process properties as well as product properties. To describe these processes and the influencing properties a process model was developed in the CRC 805. This section briefly introduces this model. Further information is given in [2, 5]. Its main components are states, processes and influencing variables. The states represent the input respectively the output of a process and are described by state matrices containing properties and a description of their scattering. A reasonable amount of the state properties refers to product properties. Processes in this model are described by the process activity, control, an operator and an identifier. Influencing properties are classified as disturbance, information, resources and user. Figure 1 depicts the notation of the process model and its elements. The combination of several processes in this notation creates a process chain.

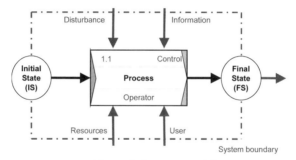

Figure 1: Process model

Semantic Technology

Ontologies are regarded as "formal models of selected aspects of the real world" [6]. They are used to represent things and their relations in a machine processable manner. Ontologies allow the assignment of higher order semantics that render the possibility of knowledge representation. The foundation of ontology languages builds the "Resource Description Framework" (RDF). It is used to represent triples of the following type:

ns:drilling   *ns:hasDisturbance*   *ns:temperature*;

This notation is called *qnames*. It is used in XML for the description of URI (Uniform Resource Identifier) references. The part in front of the colon in each element, e. g. *ns*, symbolizes the namespace which is an abbreviation for a unique URI to identify the object unambiguously. This namespace derives from the roots of RDF being a web technology. Here it is used for a further description of the element. A triple consists of subject, predicate and object. The subject, the first element of the triple, denominates the element to which the following predicate and object belong.
The predicate is the asserted property of the subject and the object builds the argument of the predicate. From these triples a graph structure emerges which allows the structured representation of knowledge. The Web Ontology Language (OWL) and its successor OWL2 are standardized ontology languages with higher expressiveness than RDF.

The basic elements of OWL are "Classes" and "Individuals" as instances of the class, as well as "Object-" and "Datatype-Properties" which allow to create relationships between Individuals and Data or Individuals and Individuals. The possibility of representing a process chain in an ontology-based information model is described in [7]. This approach uses the process model, described in the previous chapter, for collecting and structuring uncertain properties according to the processes where they are realized. The referencing of the uncertain properties to the product is also part of this ontology. The topologic structure of the products BREP (Boundary REPresentation) model is used to locate the property on the product model.

Commonly, ontology systems define binary relations between two entities. For additional attributes which describe a relation n-ary relations are used. So it is possible to append additional information to a relation e. g. source or level of trust. N-ary relations allow to state that an experiment (or person) implies that there is a relationship between entities of a knowledge base. It is possible to represent that this must not be true or that it has a general character. An n-ary relation can state that someone makes (or an experiment implies) a statement about something else without adding this result directly to the knowledge base.

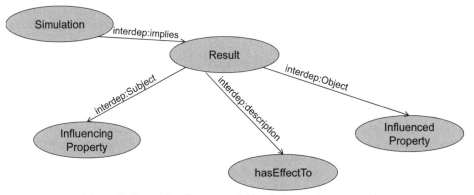

Figure 2: Example of a n-ary relation between two properties

Figure 2 shows an example of an n-ary relation. An experiment implies that an *Influencing Property* has some effect to an *Influenced Property*, and that these results seem to be trustworthy. Within a binary relation it would be:

    ns:InfluencingProperty   *ns:hasEffectTo*   *ns:InfluencedProperty.*

This relation has no additional information about source, levels of trust, which are important information for further reasoning. The availability of descriptions like *hasEffectTo* is also limited to predefined object properties. With n-ary relations more detailed descriptions are possible.

Ontology-based process and product reference

Because uncertainty occurs in processes, a crucial element of uncertainty representation is the modeling of the underlying process chains. An ontology based representation of the CRC 805 process model is presented in [7]. It is used to represent processes in the basic process model described before and constitutes the core of the uncertainty representation. The value of a property either describes the set value or an actual value resulting from a process or the processes environment.

The state properties, which mostly are also product properties, are linked to a product model such as a 3D-CAD model. For this referencing the topologic structure of the product model is used. The process properties without direct relation to the product model like influencing properties are related to the process. In this structure a set of basic elements could be identified regarding to product respectively to the process chain. Table 1 illustrates these elements according to product and process chain. Some examples are given in brackets. The first four elements of the product are used to reference product properties to the product model. Component, feature and topology elements are well-known elements in the field of 3D-CAD. The design zone is an accumulation of geometry elements which allows the locating of product properties more precisely.

The products properties build the bridge between the product and the process chain. Thus, the state properties describe the product properties in the course of the product life cycle.

The elements of the process chain correspond to CRC 805 process model. With the definition of these elements it is possible to reference uncertainty to the according subject. The values of product and process properties own a separate description of their uncertainty by the XML based uncertainty description (UDT). Objects with this description are written in *italic* letters. The description of the uncertain process property value is realized using the UDT and builds the first part of the uncertainty description. These elements aggregate to a network in the process chain as well as through the modeling of interdependencies between the different types of properties. The interdependencies of properties are important for the representation of the effects of uncertainty and will build the second part of uncertainty description in this approach.

Table 1: Basic elements of the information model

Product		Process Chain	
Component	(Connecting Element)	Process	(Drilling)
Feature	(Hole)	State	(S2-after Drilling)
Design Zone	(Shell Hole – Shell Leg)	*State Property*	*(Diameter)*
Topology Element	(Face of Hole)	*Influencing Property*	*(Temperature)*
(Product)-Property	*(Diameter)*		

The Tripod

The tripod system presented in [2] was developed as a practice example for investigations on transmission and linkage of uncertainties through different types of processes. Main purpose of this case of study was the application of some methods used for modeling and analyzing uncertainty on the example of a simple load carrying structure. For this, production and usage processes of the tripod system were considered. The tripod design was selected in order to achieve a uniform load distribution between the three tripod's legs.

Indeed, load distribution in load carrying structures was considered as an important design issue.

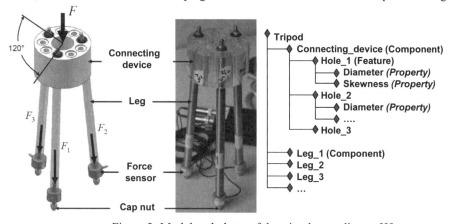

Figure 3: Model and photo of the tripod according to [2]

Therefore it is an important constraint in production and usage of a load carrying structures by the authors [2]. Figure 3 depicts the tripod and its product structure for referencing of uncertainties. The tripod consists of a connecting device and three identical legs. The leg diameters set value and its actual value is 15.0 mm.

Force cells measure the static axial force induced into each leg by an external load applied on the top surface of the connecting device. Two geometrical properties of the holes in the connecting device were considered as uncertainty sources resulting from the manufacturing process: the diameter and axis skewness β of the holes. Both uncertainty sources are inserted in the tripod with disproportionate values that exceed their real achievable values in order to increase their effects on the load distribution within the tripod.

Manufacturing of the tripod

The first example of use for the ontology-based information model deals with two connecting devices, which were manufactured for the performance of experiments. Each connecting device contains nine mounting holes, split into three groups of holes. Each group has a certain diameter. The first type of connecting devices holes has a diameter of 15.0 mm according to the set value. The others have actual values of 15.2 mm respectively 15.4 mm to simulate deviation from the set value. The first connecting device has an axis skewness of 0°deg which correspondents to the set value. A second connecting device has an actual skewness of 5°deg.

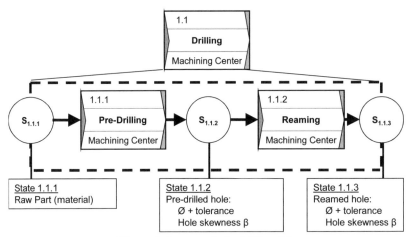

Figure 4: Process chain drilling [2]

The aim of the listed hole and skewness variation is to simulate the effect of process deviations during the manufacturing. The chosen uncertainties follow from typical failures in the drilling process. The deviation of the diameter could be caused by geometric errors of the tool which result from grinding errors in the tool production. The reason for the deviation of the skewness typically is a strength gradient of the work piece material. This effect leads to unequal cutting forces at the cutting edges of the tool and as a consequence to the holes skewness. Figure 4 illustrates the process chain drilling. It is a two-step process consisting of pre-drilling and reaming. The result of the pre-drilling process is a hole which is described by the state $S_{1.1.2}$. Afterwards a reaming process is made for the hole finishing. The final mounting hole is described by the properties specified in the final state.

Assembly and load distribution of the tripod

The second example deals with the usage of the tripod system. Figure 5 shows the analyzed overall process chain for the tripod system, consisting of three main processes: drilling, assembly and load distribution [2]. The states of the process chain contain the data regarding material (S_1), the geometric parameters of the holes and the legs geometric properties (S_2, S_3) as well as the resulting load distribution (S_4).

As mentioned above, the final state of the drilling process is the reamed mounting hole with the uncertainties diameter and skewness deviation. Thus, an eccentricity of the legs in the mounting holes results in the assembly process. This causes an unequal load distribution in the usage of the tripod.

To analyse the influence of the mounting process, the tripod is repeatedly assembled and disassembled and the resulting forces F_1, F_2 and F_3 in the legs are measured. Due to the deviations in hole diameter and angle in the connecting device, different deviations from the ideal leg position are resulting. This leads to a broad scatter of the measured leg forces. The different diameters of the holes in the connecting device give the possibility to adjust different spread of the leg position and the resulting forces. As a consequence it is possible to simulate different effects and the resulting uncertainty with only one tripod. As expected, a similar trend is shown by an analytical simulation. Therein, the variance of the measured axial forces in the legs also grows with an increasing deviation of the holes diameter. Other uncertainties like the sensitivity of the force sensors, the clearance fit between leg and hole, varying length of the legs due to tolerances in manufacturing process etc. have been recognized as influencing factors on the load distribution in tripod. Therefore, further studies for the tripod are made to identify and classify all possible uncertainties influencing the load distribution. The knowledge about the process properties and their deviations, e. g. manufacturing deviations like tool geometry errors or strength gradients in the material, can be crucial for the product development. Once the exactness of the hole is identified as important, it is necessary to gather information about the processes and the interdependencies between them.

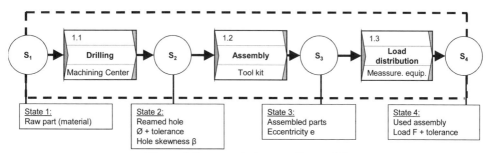

Figure 5: Process chain according to [2]

For the representation of this knowledge, an ontology based information model is developed and demonstrated using the example of the tripod system. The standardized information model allows exchanging information about uncertainty and comparing these uncertainties along the product life cycle.

A new approach for the exchange of uncertainty

As mentioned before, uncertainty consists of two elements, uncertain properties with different types of description, and the effect of the scattering of one property to other properties. Within the tripod experiments, several researchers from different institutes were involved. Thus, there is the need for an exchange of data describing the different levels of uncertainty similar to real product development. Three main aims for the exchange of uncertainty between different stakeholders are identified:

- Allocation and description of uncertain properties in product and process
- Description of uncertain property values
- Representation of interdependencies identified in the experiments

The description of uncertain property values is realized by using the UDT. This is implemented by connectors for the deployed software, providing a common description of uncertain properties. In this case UDT-XML files describe histograms from the load distribution in the legs as well as geometry properties describing the drilled hole. The storage of further statistic indicators is possible. The generated UDT-XML files are integrated into the ontology process model as "Datatype Properties" according to the described property. This allows the direct integration of the uncertain property description into the ontology. In general, for the data handling, especially for extensive data sets, it has to be clear to which leg and experiment a load distribution belongs and which state of the manufacturing process is described. This is realized by the ontology process model containing the product properties and structure in the state properties. By this means every participant is able to supplement information to and receive information from the system. The possibility of splitting the ontology in several parts is an advantage of this approach. Each stakeholder will be able to work with core ontology and supplement further information. These ontologies then can be imported into a common knowledge base which renders the possibility of distributed work. The core ontology contains the process model with the used state properties that are connecting the processes.

While the scattering property can be described by UDT-XML files there is also a need for a cause and effect relation in the information model. An obvious way to model this effect would be a simple object property in the ontology. This has several shortcomings. The interdependencies could not be supplemented by a further description like e. g. trustworthiness, source or further quantification.

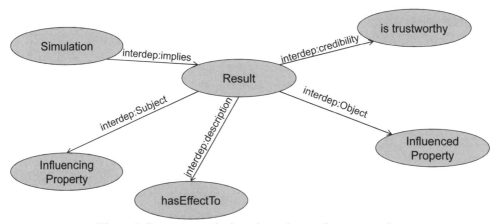

Figure 6: Statement to the interdependency of two properties

However, this information is important for the evaluation of the interdependencies. Furthermore, the results of an experiment can be vague and reflect only assumptions. This information is also important for a cause and effect relation. The approach, described in this paper, uses n-ary relations in combination with an object property which indicates the possible way of influence (or non-influence). This complies with the proposal of the World Wide Web Consortium (W3C) OWL working group for negative and axiom annotations [8]. The use of n-ary relation allows supplementing further information to such statements. This information enables describing the interdependencies and renders the possibility for further classification and reasoning with this new type of element in the process model. The used approach gives the ability to make statements about two properties. Making statements about groups of properties will be part of further research work.

Figure 6 shows such reification with additional information. The statement contains the information that an experiment implies, that *Influencing Property* has an effect to *Influenced Property*. Further description of the influence can be given to the *hasEffectTo* element.

Also an example for the credibility of this experiment is given. Table 2 shows a short presentation of the triples used for the n-ary relation. The used namespaces (*xyz:*) comply with the type of element and are derived from the elements name or are used to group the elements. The definition of uncertain properties and the referencing to product and process allows the collaboration and exchange of information about them. This information is represented by the UDT respectively by the n-ary relation.

Table 2: Example of a n-ary relation of an interdependency

organisation:experiment_1	interdep:implies	result:experiment
result:experiment	interdep:Subject	process:InfluencingProperty
result:experiment	interdep:description	interdep:hasEffectTo;
result:experiment	interdep:Object	process:InfluencedProperty
result:experiment	interdep:credibility	interdep:isTrustworthy

Knowledge from the drilling experiments

The focus of the experiments for the process chain drilling and reaming are the diameter accuracy and the radial deviation. These are important properties for the function of the tripod as well as the evaluation of resulting hole quality. Experiments were made for the identification of the connection between a pre-drilled hole skewness and the reamed hole. The analysis of the results shows that the common assumption that the reamer follows the pre-drilled hole is not correct for the chosen tool diameter and L/D-ratio. This knowledge is important for further analysis and is represented by a statement in table 3.

Table 3: Knowledge from drilling experiments

ptw:drillingExperiment	interdep:implies	result:experiment_1
result:experiment_1	interdep:Subject	stateProperty:PreDrilledHoleSkewness
result:experiment_1	interdep:description	interdep:hasMarginalEffectTo;
result:experiment_1	interdep:Object	stateProperty:finalHoleSkewness;
result:experiment_2	interdep:credibility	interdep:isTrustworthy

Knowledge from the load distribution experiments

The load distribution experiment focuses the deviation from an ideal distribution of 1/3 in each leg. Due to deviations in the mounting holes and the resulting eccentricity of the legs in the assembly process, deviations from this ideal are expected. The experiments confirm these expectations, thus the resulting statement is shown in Table 4.

Table 4: Knowledge from stressing experiments

szm:stressingExperiment	interdep:implies	result:experiment_2
result:experiment_2	interdep:Subject	stateProperty:finalHoleSkewness
result:experiment_2	interdep:description	interdep:hasRecognizableEffectTo
result:experiment_2	interdep:Object	stateProperty:LoadDistributionL1
result:experiment_2	interdep:credibility	interdep:isTrustworthy

Visualization of results from experiments

The core of the visualization concept is the bidirectional associativity between uncertain information visualized in the uncertainty-browser and the parameters presented in a CAD-system. This potential allows a variation of geometry in the lowest and highest border of the given value of deviation.

Basically two possibilities to visualize knowledge from the presented experiments exist. Each possibility uses the front-end of an uncertainty-browser presented in [9, 10]. The uncertainty-browser contains the interpreted information about uncertain property, the according process and the allocation within the geometry model from the information model. The geometry references are imported along the CAD-model structure and adapted with the ontology-based system [10]. For this the BRep structure is used.

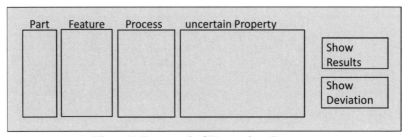

Figure 7: Front-end of Uncertainty-Browser

Every uncertain property owns information about the known deviation, the value of the deviation and its results. These results and the values of deviation will be shown separately on the geometry by confirming the "show-buttons" in the uncertainty-browser. The user has the possibility to choose between two visualization-views relating their content (shown in Figure 7): Show results or show deviation.

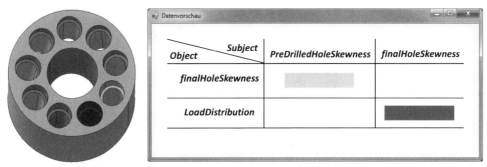

Figure 8: Visualization of results on a 3D-CAD geometry

By confirming the "show results-button" the results describing the interdependencies between states will be visualized in a separate window. Complex coherences resulting out of the experiments can be shown. The matrix (Figure 8) provides a presentation for the queried information respective their uncertain property and theirs interdependencies. The information about the amount of the interdependencies and their effect is realized by a color scale. In this example the presentation of a marginal effect is achieved by a less intensive color, a recognizable effect in contrast is realized by a strong color.

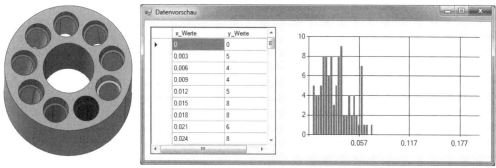

Figure 9: Visualization of information about uncertainty (deviation) on 3D-CAD geometry

Second possibility to visualize knowledge from experiments is to get information about the deviation of uncertainty and the uncertain property. This will be achieved by selection of the "show deviation-button". This selection causes a query of information related to the geometry and its deviation. Result is a presentation of the UDT and its data. In this case a histogram describes the deviation of the drilled hole skewness. Furthermore the uncertain property skewness of the hole" is connected with a parameter within the CAD-system. Bidirectional associativity of parameters between the uncertain property and the CAD-system allows a variation of geometry in the range between the lowest and highest border of the given value of the deviation.

Conclusion

This paper introduces an approach for the representation of uncertainties in load carrying structures. The handling of uncertainty is a problem which affects the whole product life cycle and different stakeholders. Thus a demand for the exchange of knowledge about uncertainty emerges. This approach gives the ability to represent the uncertain process property as well as cause and effect relations in an ontology based information model. This enables the possibility for the distributed capturing and exchange of uncertainty related information. A case study with a simple load carrying system, including manufacturing and usage, was conducted and presented. Besides the exchange of uncertainty the presentation to the designer is very important. An approach for the visualization is also given in this paper.

Acknowledgments

We like to thank the German Research Foundation (Deutsche Forschungsgemeinschaft - DFG) for funding this project within the Collaborative Research Centre (CRC) 805.

References

[1] Hanselka, H.; Platz, R.: *Ansätze und Maßnahmen zur Beherrschung von Unsicherheit in lasttragenden Systemen des Maschinenbaus (engl.: Controlling Uncertainties in Load Carrying Systems)*. Konstruktion 11/12 2010, 2010.

[2] Platz, R.; Ondoua, S.; Habermehl, K.; Bedarff, T.; Hauer, T.; Schmitt, S. & Hanselka, H.: *Approach to validate the influences of uncertainties in manufacturing on using load-carrying structures*. In: Proceedings of the International Conference on Noise and Vibration Engineering ISMA 2010, 2010.

[3] Engelhardt, R.A.; Koenen, J.F.; Enss, G.C.; Sichau, A.; Platz, R.; Kloberdanz, H.; Birkhofer, H.; Hanselka, H.: 2010. *A Model to Categorize Uncertainty in Load-Carrying Systems.* In Proceedings of the 1st International Conference on Modeling and Management of Engineering Processes. Springer Verlag, 2010.

[4] Agarwal., H.; Renaud, J. E; Preston, E. L.; Padmanabhan, D.: *Uncertainty quantification using evidence theory in multidisciplinary design optimisation.* In: J. Reliability Enginering and System Safety. 85. 281-294, 2004.

[5] Sprenger, A.; Mosch, L.; Mecke, K.; Anderl, R.: *Representation of Uncertainty in Distributed Product Development,* 17th European Concurrent Engineering Conf. ECEC2011, 2011.

[6] Gruber, T.: *A Translation Approach to Portable Ontology Specifications.* Knowledge Acquisition Academic Press Inc, 5(2), 1993.

[7] Mecke, K.; Sprenger, A.; Mosch, L.; Anderl, R.: *Beherrschung von Unsicherheit von lasttragenden Systemen im Maschinenbau durch Ontologie-basierte Informationsverarbeitung und Visualisierungstechnik (engl.: Control of Uncertainties in Load Carrying Systems by ontology based information processing and visualisation).* In KT 2010 - Kolloquium Konstruktionstechnik, Magdeburg, 2010.

[8] Defining N-ary Relations on the Semantic Web, *W3C recommendation 12 April 2006,* http://www.w3.org/TR/swbp-n-aryRelations/.

[9] Mosch, L.; Sprenger, A.; Anderl, R.: *Approach for Visualization of Uncertainty in CAD-Systems based on Ontologies.* In: Proceedings of the ASME 2010 International Mechanical Engineering Congress & Exposition (IMECE).

[10] Mosch, L.; Sprenger, A.; Anderl, R.: *Consideration of Uncertainty in Virtual Product Design.* To appear in Proceedings of the ASME 2011 International Design Engineering Technical Conferences and Computers and Information in Engineering Conference (IDETC/CIE 2011).

An approach of design methodology and tolerance optimization in the early development stage to achieve robust systems

Philipp Steinle[1,a], Martin Bohn [1,b]

[1]Daimler AG, Benz-Straße, 71063 Sindelfingen, Germany

[a]philipp.steinle@daimler.com, [b]martin.bohn@daimler.com

Keywords: Tolerance management, robust design, application of design methods

Abstract: In the early stages of the vehicle development basic decision for the future vehicle concepts are made. Anyhow not all information for a confirmed decision, especially in the field of tolerance management, is available at that time. Known methods to estimate the impact on the final product normally require determined concepts and nearly completed parts and are thus not feasible to find new solutions. One solution to overcome this problem is the consequent use of design methodology. Combined with tolerance aspects the result of the process is a robust concept for all following process steps.

Introduction

The automotive industry tries to reduce the development costs and the time needed to bring a car to market by shortening the vehicle development. For this process several methods which formally describe the sequence from the early development stages with the problem analysis up to the documentation exist, e.g. VDI2221 [5]. Regardless of the characteristic of the process either as a set of cycles to go through or a process with overlapping phases, all equal in the fact that fundamental decisions influencing the result of the process are made at an early concept stage. At that time of the development process normally not all relevant facts are known to evaluate the system. Basis of these decisions must not exclusively be experience but logic and replaceable procedures within a company.

Scientific theses regarding uncertainties in the product development provide an overview about different methods to deal with unknown or not precisely known parameters, compare [9], [10]. These methods help to estimate and reduce the risk on the development process resulting from the uncertainty and subsequently rise the quality of the decisions.

A field of engineering normally not connect with the early stage of development is the tolerance management. However, the tolerance management as a cross-sectional discipline is among other fields of engineering responsible for the costs, the developing time and the quality of the product [3]. Beyond this, tolerance management also influences important functional characteristics of the vehicle.

In the development process tolerance management as well as robust design strategies are not commonly used within the early process stage, as they relay on given concepts and designed part geometries. Therefore they are normally used at a later stage to the development process at which the system can only be slightly adapted.

Tolerance management as part of robust design

Every produced part or assembly is subject to deviations from the original defined geometry. Thereby the complexity to connect parts and assemblies rises with the number of them. The result of the process (vehicle) however must ensure a high dimensional stability. Thus the automotive industry needs a holistic process to evaluate single parts and assemblies. The basis therefore is a complete and consistent description of the datums and their orientation as well as a function orientated tolerance method. [7]

Out of the tolerance analysis process, which requires measurable and testable tolerances, parameters can be identified and means for an improvement can be exposed. These results influence the next development process to steadily improve the products.

Tolerance management aims to set the tolerances as large as possible and as small as necessary. Combined with the idea of robust design, which is according to Taguchi "the state where the technology, product, or process performance is minimally sensitive to factors causing variability" [8] the quality of a product can be increased with only minor influences of the production costs.

Mostly these optimizations take place, when the principal concept of the system is determined and the parts are already designed [2]. Afterwards numerous optimization runs according a Design of Experiment-strategy can be conducted. Thereby software tools like OptiSlang automatically combine the different input parameters and control the simulations [6].

Based on the necessity of concrete concepts and boundary conditions for the optimization robust design tools and tolerance simulation software can hardly be used in the early stage of product development. Therefore processes must be defined to use the idea of a robust tolerance design as early as possible.

Method to reduce the influence of deviations in the early development stages

The aim of the process is to find basic solutions for existing mechanical problems by means of the methodological design with the focus on tolerance management. The solutions should offer a robust system behavior against the variation to the input parameters.

Fig. 1 shows the generalized process to optimize the mechanical system. The current technical problem is the input for the process. Additionally the experience of previously build vehicles is added. Within the design methodology, the technical system is abstracted, if necessary in several loops. Subsequently the problem can be split up in different partial problems. Thus it is easier to find solutions which, in a next step, can be combined to represent the full system. These abstracted mechanical systems were evaluated under the aspect of tolerance management. Thus a tolerance robust design represents the end of the process with the output of the results.

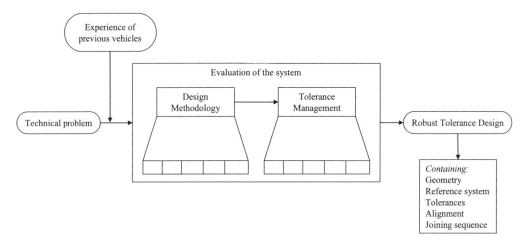

Fig. 1 General process from the input of the data, over the system evaluation with the means of design methodologies and the tolerance management till the output of the results.

The methodological design is subdivided into different steps. In a first step the system must be abstracted in several loops to filter the basic functions of the system. This leads to a much better understanding of the system and it's interactions. Afterwards the global problem can be split up in several, individual solvable problem definitions. In case of a shaft, the system's function is to forward a torque and bear forces. For a static defined support of the shaft two bearings are necessary. Fig. 2 shows the abstraction of a shaft with the bearings and their position relative to the shaft. In case of the shaft, there are several solutions to solve the problem, e.g. different kinds of bearings which and have to be chosen according to the operating conditions. Normally not all solutions can be combined as some of them preclude each other. Thereby the number of systems can be reduced.

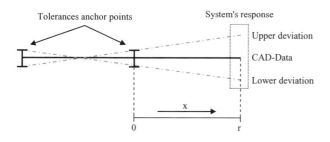 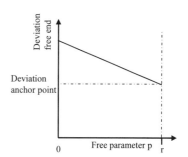

Fig. 2 Simplified tolerance model with two supporting points, their deviations and the spread of the system's response

Fig. 3 Illustration of the free end's deviation in dependence on the parameter p and the restriction r

The tolerances of the system are composed of the tolerance of the supports and the geometric effect of the part or the assembly. Leverage effects can thereby diminish or multiply the effect of tolerances. Input parameters for the tolerance consideration are the position of the part(s), their orientation within the design space and their connection. Next to the tolerances of the parts in there must be at least one free parameter p to optimize the system. Beyond this it can be required to restrict the optimization process to a given solution space r. According to the dimension of the problem a suitable optimization strategy must be selected [4]. In case of the small example in Fig. 2 a one dimensional optimization strategy is sufficient. Therefore the output of the process can be illustrated as a graph with the deviation of the free end over the parameter x restricted trough the length of the part at the position r, compare Fig. 3.

Case Study: Tailgate

The previous introduced method will now be adapted to a complex example of the automobile industry. As a case study for a tolerance influenced robust design serves the opening of a tailgate. Customer relevant geometrical parameters are for instance the height of the vehicle with opened tailgate and the height beneath the tailgate.

The abstraction of the system leads to a mechanical system with two simplified parts (representing the body and the tailgate) and a pivot point between them. The functional description of the given problem is to generate a torque around the hinge of the tailgate for opening and closing. This torque can result from a force with a given distance between the attachment points and the hinge. The further subdivision for one possible path is illustrated in Fig. 4.

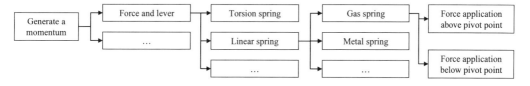

Fig. 4 Chose of the gas spring and the position within the vehicle

Looking at the tailgate the possible solutions can be grouped into two separate fields of solutions. The distinguishing factor is the effective lever of the vector caused by the force of the gas spring and the pivot point of the tailgate in it's opened position. An imaginary horizontal line through the pivot point divides the two solution spaces in an upper and lower one, see Fig. 5 and Fig. 6.

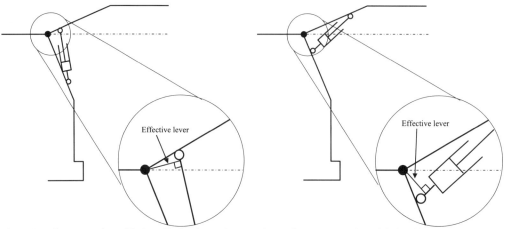

Fig. 5 Standing gas spring with the effective lever above the hinge (System A)

Fig. 6 Flip over gas spring with the effective lever beneath the hinge (System B)

The attachment points of the gas spring and the spring itself in it's end position are subject to tolerances, compare Table 1. Not considered are the tolerance of the hinge, the tolerance of the car body and the tolerance of the chassis as these tolerances are the same for each possible mounting position of the gas spring. Table 1 contains the responding tolerance of the lower edge of the tailgate (here the most right point of the tailgate in the opened position) according to the tolerances of the system for a standing and a flip over position of the spring under the assumption of equal tolerances for both solutions.

Table 1 Tolerances of two solutions of the robust design process

	Standing gas Spring	Flip over gas spring
Tolerance upper attachment point (tailgate) [mm]	± 1,0	± 1,0
Tolerance lower attachment piont (tailgate) [mm]	± 1,5	± 1,5
Tolerance gas spring [mm]	± 2,0	± 2,0
Tolerance of the lower edge of the tailgate [mm]	± 17	± 25

Based on the geometrical proportions the tolerance of the tolerance chain is multiplied especially by the ratio of the distance from the hinge to the attachment point on the tailgate compared with the distance from the attachment point to the lower end of the tailgate (on the pictures the rightmost point of the lid). System A in this case depends far more on tolerances of the input parameters compared to system B. From the tolerance management's point of view system B would therefore be the favorite.

As every product there is always a compromise between numerous requirements, therefore both variants are currently used in the automotive industry.

In a further process step the supports of the spring can be varied in two directions. Restrictions in this example are the vehicle design, the design space and the maximum length of the gas spring, compare [1].

Conclusion

Tolerance management as part of robust design represents an important step in the vehicle development process. However, special tools for dealing with either tolerance management or robust design normally base on already designed parts and concept. In the early stage of the development processes new methods must be found to integrate the idea behind robust tolerances. By means of design methodologies the system can thereby be transformed into an abstract model which is the basis for the further processes. By solving the abstract model and the composition of new solutions, the whole system can be evaluated under the aspect of a tolerance robust design. As an example the tailgate of a vehicle was used to demonstrate the method and their effects on the tolerances.

References

[1] M. Bohn, F. Wuttke: Optimization of robustness as contribution to early design validation of kinematically-dominated mechatronic systems regarding automotive needs, NAFMES Nordic Regional Conference, Gotheburg, 2010

[2] G. Wehr: Kopplung von stochastischer FEA und Schädigungsanalyse (engl.: Coupling of stochastic FEA and damage analysis), Weimarer Optimierungs- und Stochastiktage 2.0, 2005

[3] Bohn, M.: Toleranzmanagement im Entwicklungsprozess. Reduzierung der Auswirkungen von Toleranzen auf Zusammenbauten der Automobil-Kraosserien (engl.: Tolerance management in the development process. Reduction of the effects of tolerances on assemblies of vehicle bodies), Dissertation, Universität Karlsruhe. 1998

[4] L. Harzheim: Strukturoptimierung, Grundlagen und Anwendungen (engl.: Structure optimization, fundamentals and applications), Verlag Harry Deutsch, 1. Auflage 2008

[5] VDI-Richtlinie 2221: Methodik zum Entwickeln und Konstruieren technischer Systeme und Produkte (engl.: Methodology to develop and design technical systems and products), Beuth Verlag, 1993

[6] P. Gust, C. Schluer: Robustes Design versus Toleranzmanagement am Beispiel eines KFZ Schließsystems (engl.: Robust design vs. tolerance management by taking the example of a automobile closure system), Weimarer Optimierungs- und Stochastiktage tage 6.0, Weimar, 2009

[7] K. Hetsch: Toleranzmanagement - Methodik Form- und Lage (engl.: Tolerance management – Methodology, form and position), Vorlesung FH Aachen, 2009

[8] G. Taguchi, S. Chowdhury, S. Taguchi: Robust Engineering. McGraw Hill, New York, 2000

[9] M. Lorenz: Handling of Strategic Uncertainties in Integrated Product Development, Dissertation, Universität München, 2008

[10] S. Grunwald: Methode zur Anwendung der flexiblen integrierten Produktentwicklung und Montageplanung. (engl.: Method for the application of a flexible product development and assembly planning), Berichte aus dem Institut für Werkzeugmaschinen und Betriebswissenschaften der technischen Universität München, Band 159, Herbert Utz Verlag GmbH, 2002

2. Uncertainty in Production

Influence of Tolerances on Mechatronic Comfort Systems Behavior

Martin Bohn[1,a] and Fabian Wuttke[1,b]

[1]Daimler AG, Benzstraße, 71063 Sindelfingen, Germany

[a]Martin.Bohn@daimler.com, [b]Fabian.Wuttke@daimler.com

Keywords: Robust Design, Evaluation of Robustness, Mechatronic Systems, Comfort Systems.

Abstract. In contrast to Robust Design applications for FEM-modeled parts, the simulation of a mechatronic system including both mechanical and electrical parts requires a different strategy for the investigation of its robustness. Differences mainly results from interactions between the mechanical and the electrical part of the mechatronic system. Furthermore, a comfort mechatronic system in the automotive industry is designed customer-oriented. Subsequently, the behavior of its movement and the respective robustness has to be considered too. This paper presents an approach to evaluate robustness of mechatronic comfort systems. Additionally, the approach is applied to sliding doors of vans to prove the practicability to industrial problems.

Introduction

In order to improve the customer's comfort, many modules in the automotive industry have shifted from solely mechanical solutions to mechatronic systems. In the past, customers had to use their hands and hence put their own energy into the change of state of the module. As customers do not only link luxury of cars to expensive interior, many automotive companies put R&D effort into the development of comfort features. Those features aim to make life easier for the customer and to help companies to differentiate from the competitors [1]. Given examples are window lifters, retractable hardtops and automatic doors (tailgates, sliding doors, etc.). During the development of comfort modules, engineers seek to find an optimum within the classic development triangle of time, cost and quality [2]. Supported by current discussions about CO_2 reduction which leads to lightweight car design [3], this triangle is more and more extended to a rectangle by the factor weight, see fig. 1.

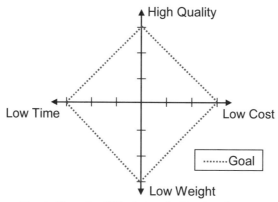

Fig. 1: Targets of Mechatronic Comfort Systems

Hence, the goal of a mechatronic automotive comfort system is to have a simultaneously optimized module. The goal of minimum time is most often a question of efficient and effective development processes. But simultaneously minimizing the targets cost and weight is almost never obtainable. This is because reaching the wanted quality while becoming lighter cannot be done accompanied by saving money. Lightweight design in terms of substitution of materials (aluminum alloys instead of steel) or concepts (function integration of casting parts instead of welded parts assembled by many single components) is a cost driver. Therefore, the parts of many comfort modules represent a compromise between costs and weight.

Additionally, lightweight-optimized parts tend to become vulnerable to uncertainties during production and use - they lose robustness and / or reliability [4]. This is shown in fig. 2, illustrating different output characteristics of a conventional and a lightweight part design.

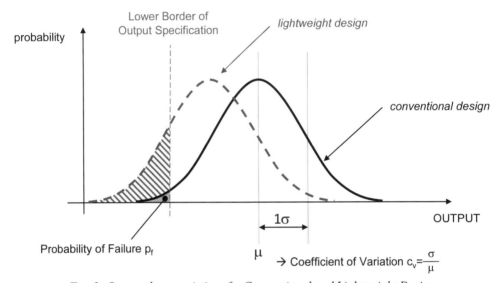

Fig. 2: Output characteristics of a Conventional and Lightweight Design

Adopted from the target pattern in fig. 1, quality Q is translated to robustness as Coefficient of Variation c_v and reliability as Probability of Failure p_f in this paper. Taking into account the number of functional requirements m, the quality of the product can be shaped as

$$Q = \frac{1}{\sum_{i=1}^{m}(\frac{a_i \cdot c_{v,i} + b_i \cdot p_{f,i}}{a_i + b_i})} \in (0,\infty), \tag{1}$$

with a_i and b_i representing weighting factors between robustness and reliability, respectively. Q is a result of uncertainties and should be as high as possible. Q can be interpreted as the inverse of the weighted average of c_V and p_f for each functional requirement. Therefore, Q^{-1}/m can be used to identify the average robustness of all functional requirements. Q^{-1}/m should be well below 0.05 to ensure a robust system behavior. Uncertainties usually come from different part designs within their productions tolerances or from different use behavior of the customer. For mechatronic comfort systems, questions on different levels remain open:

1. Influence of uncertainties of production and use on system behavior
2. Contribution of the mechanical and electrical part of uncertainties on system behavior
3. Significant uncertainties.

Following, an approach is proposed to address these questions.

An Approach to Evaluate Robustness of Mechatronic Comfort Systems Behavior

When investigating the robustness of a product, the so call P-diagram is often taken into account [5]. P-diagrams visualize the relation between the product outputs and the influencing factors. Figure 3 shows a P-diagram applied to mechatronic comfort systems.

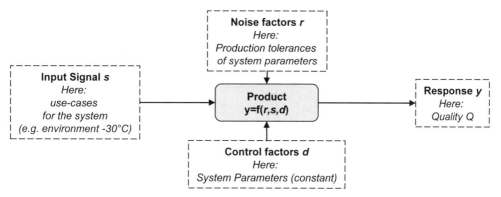

Fig. 3: P-Diagram of Mechatronic Comfort Systems

According to Figure 3, the product's quality Q depends on 3 influencing factors. Mechatronic comfort systems in the automotive industry have to work under very different external conditions. The claim of stable performance all over the world is typically written down in requirement specifications. The different conditions are typically checked by use-cases which are described here as input signals s. Therefore, mechatronic comfort systems have to be classified as dynamic systems according to Taguchi [6]. Hence, uncertainties of use can be separated from uncertainties of production, because it's sufficient to test a product under extreme conditions – classified as input signals s. All uncertainties coming from different production conditions are put into noise factors r which influence the system's behavior randomly in reality. In the framework of P-Diagrams, control factors are typically included. In this paper, emphasis is on the investigation of a module, not on the design process of the module. Therefore, control factors d remain constant.

The Quality of Behavior. In contrast to most other mechanical or mechatronic systems, a comfort system's functional requirements not only focus on the fulfillment of strict functional requirements regarding performance (e.g. opening times) or applied loads. Rather, the behavior of the entire system is the most important investigation object. Thus, the dimensions of robustness evaluation are extended in order to determine robustness considering kinematic parameters like opening stroke, see fig. 4. Consequently, not only statistical values of one output probability density function are examined but also the behavior of these statistical values plotted against the kinematic parameters [7].

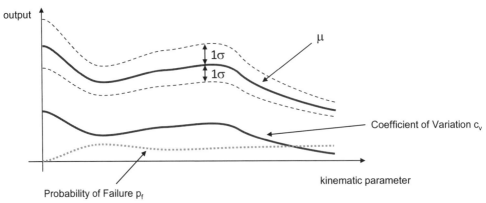

Fig. 4: Dimension extension of robustness evaluation

Evaluation Process. The quality Q of the mechatronic system is composed of the qualities Q_i of every use case i. This means that a sufficient number of samples needs to be created for every use-case i. According to equation (1), Q_i can then be evaluated based on the Coefficient of Variation c_v and the Probability of Failure p_f. Beyond a number for Q_i, engineers are more interested in the specific production tolerances which cause Q_i. Hence, the correlation between production tolerances and the output associated to Q_i has to be determined. This can be done by statistically identifying the coefficients of correlation ρ between every production uncertainty and Q_i. Coefficients of correlation can be calculated for different regression assumption, e.g. linear or quadratic regression. Assuming a well-known system with small deviations of production tolerances, the coefficient of correlation is built linearly after Bravais-Pearson [8]

$$\rho_{A,B} = \frac{1}{n-1} \frac{\sum_{k=1}^{n}(A_k - \mu_A)(B_k - \mu_B)}{\sigma_A \sigma_B} \qquad (2)$$

n represent the number of calculated samples. A_K and B_K describe the values for each point, whereas μ_A and μ_B demonstrate the corresponding values of the regression function. σ_A and σ_B specify respective standard deviations. The absolute value of the Coefficient of Correlation $\rho_{A,B}$ is between 0 (no correlation) and 1 (perfect correlation).

In order to prove that the deviations of the output result from deviating inputs, the Coefficient of Determination (COD) is often taken into account [9].

$$COD = \frac{\sum_{i=1}^{n}(\hat{y}_i - \bar{y})^2}{\sum_{i=1}^{n}(y_i - \bar{y})^2}$$

$\hat{y}_i$ represents estimated values of the output, actually measured of values of the output are represented by $\bar{y}_i$. $\bar{y}$ indicates the total mean value of the output. If the Coefficients of Determination (COD_i) for every output associated to Q_i meet the wanted minimum level of explanation (typically $COD_i > 0.2$), the different levels for the coefficients of correlation sufficiently illustrate the relationship between r and Q_i. As it can be seen in fig. 5, this approach enables a deep and concrete analysis of how production tolerances influence the behavior of mechatronic comfort systems.

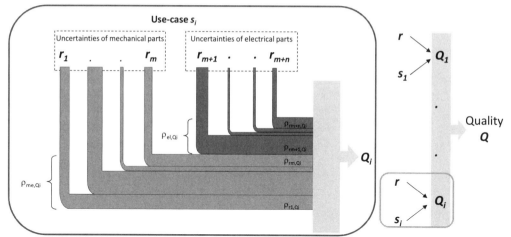

Fig. 5: Visualization of the Influence of Production Tolerances on System Quality Q

First, the important production tolerances can be determined for every use-case s_i by defining significance borders for a minimum Coefficient of Correlation. Second, production tolerances and their respective Coefficients of Correlation can be structured into the aggregated influence of mechanical parts $\rho_{me,Qi}$ (e.g. tolerance of position) and electrical parts $\rho_{el,Qi}$ (e.g. torque deviation of electrical drive unit). Thereby, problems of mechatronic systems can be traced back to the important mechanical or electrical components.

Example

Mechatronic comfort systems do not only spread to the private sector but also to business applications like delivery vans. Popular examples are automatic sliding doors. They help to save time during delivery because the driver doesn't any longer need to close or open the sliding doors manually. Typical systems contain two or three sliding rails at the door frame or at the inner side of the door. Track carriages containing rolls on the opposite side allow the translational movement. The actuator usually drives the track carriages or its rolls by a Bowden cable.

Shown in fig. 6, the P-Diagram has been applied to automatic sliding doors, separating the production uncertainties of the mechanical and electrical part from each other.

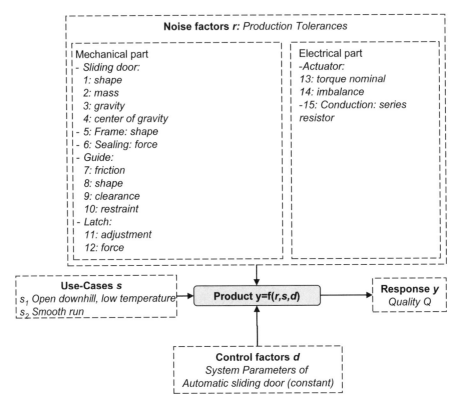

Fig. 6: P-Diagram for automatic sliding doors

Following, the simulation results for the exemplary use-cases s_1 and s_2 are shown. Simulation is done with 250 Latin Hypercube Samples describing different combinations of the production tolerances *r*. These production tolerances are described by normal distributions and specific values for µ and σ. The minimum magnitude for Coefficient of Correlation $\rho_{r,Q}$ is defined as $\rho_{r,Q}=0.2$, otherwise the production tolerance can be regarded as not correlated to the output associated quality Q.

The use-case s_1 describes a very demanding functional requirement where the automatic sliding door has to be opened downhill, which enlarges the necessary opening force compared to the horizontal case. Low temperatures far below the freezing point additionally make the opening movement even more difficult. The investigated output of s_1 is the minimum power reserve of the actuator over the opening stroke. Table 1 shows the results for use-case s_1.

Table 1: Significant production tolerances for use-case s_1

Significant production tolerance	$\rho_{ri,Q1}$
Nominal torque (el.)	0.92
Guide restraint (mech.)	-0.79
Mass of sliding door (mech.)	-0.60
Guide friction (mech.)	-0.41
Series resistor (el.)	-0.32

As it can be derived from table 1, the number of significant production tolerances can be reduced from 15 (shown in fig. 6) to 5. The most dominant influencing factor is the nominal torque of the actuator deviating around 30%. The larger the nominal torque of the actuator is, the more power reserve is available. This leads to a positive value of the Coefficient of Correlation. The remaining 4 significant production tolerances have a negative effect on the minimum power reserve. Therefore, the Coefficient of Correlation becomes negative. Investigations on a higher hierarchical level result in a relation of the mechanical and the electrical part of 3:2. Overall, engineers are supposed to put emphasis on the optimization of the mechanical part although the most dominant production tolerance is an electrical one and must not be neglected.

The use-case s_2 addresses the claim of a customer-oriented system. The movement of the sliding door has to be as smooth as possible. Therefore, the considered output of s_2 is the maximum acceleration of the door over the opening stroke after reaching the nominal velocity. This output itself has to be as small as possible. Table 2 shows the results for use-case s_2.

Table 2: Significant production tolerances for use-case s_2

Significant production tolerance	$\rho_{ri,Q2}$
Guide restraint (mech.)	0.96
Guide friction (mech.)	0.73
Guide clearance (mech.)	0.71

As it can be derived from table 1, the number of significant production tolerances can be reduced from 15 to 3. These 3 influencing factors are all mechanical ones and do have a positive effect on the maximum acceleration because the Coefficients of Correlation are all positive. Therefore, the 3 mentioned production tolerances need to be reduced. All of them are mechanical ones, the electrical part of the mechatronic comfort system does not play a major role for s_2.

Conclusion

Mechatronic systems for comfort purposes in the automotive industry become increasingly important. Accompanied by the competitive automotive environment, a detailed investigation method of the influence of tolerances on the behavior of such systems becomes inevitable. Therefore, this paper proposed an approach to trace back deviations of the comfort system's behavior to the significant production tolerances. By analyzing the coefficients of linear correlation between every production tolerance and every output, the importance of all production tolerances can be compared to each other. Additionally, it is proposed to structure tolerances of mechatronic comfort systems into a mechanical and an electrical part. Thereby, problems can be addressed on a less-detailed and higher hierarchical level. Further investigations seek to identify interaction levels between the mechanical and the electrical part as well as to bring in advanced regression models for statistical analysis.

References

[1] Sass, R. *New Again: The Hideaway Hardtop*, The New York Times, 2006, available online at http://www.nytimes.com/2006/12/10/automobiles/10RETRACT.html?ex=1323406800&en= a440f0f4ff67f836&ei=5090&partner=rssuserland&emc=rssfdsa

[2] Lindemann, U. *Methodische Entwicklung technischer Produkte (engl: Methodical development of technical products)*, 3 ed., 2009 (Springer Dordrecht Heidelberg London New York) DOI 10.1007/978-3-642-01423-9

[3] Ma, M. and Yi, H. *Lightweight Car Body and Application of High Strength Steels*, 2011. In Advanced Steels, Part 3, pp. 187-198 DOI: 10.1007/978-3-642-17665-4_20

[4] Bucher, C. *Basic concepts for robustness evaluation using stochastic analysis*. In EUROMECH Colloquium 482, September 2007

[5] Phadke, M.S. *Quality Engineering using robust Design*, 1989 (Prentice Hall, Englewood Cliffs, New Jersey)
[6] Taguchi, G. S. *Introduction to Quality Engineering – Designing Quality into Products and Processes,* 1986 (Asian productivity Organization, Tokyo)
[7] Wuttke, F. and Bohn, M. *Optimization of Robustness as Contribution to Early Design Validation of Kinematically-Dominated Mechatronic Systems Regarding Automotive Needs*. In NAFEMS Nordic Conference 2010: Future Trends and Needs in Engineering Simulation, October 2010
[8] King, B., Rosopa, P. and Minium, E. *Statistical Reasoning in the Behavioral Sciences*, 5 ed., 2007 (Wiley & Sons, Hoboken, New Jersey)
[9] Kessler, W. *Multivariate Datenanalyse in der Bio- und Prozessanalytik (engl: Multivariate data analysis in bio- and process analytics),* Wiley-VCH, 2007

Control of uncertainties in metal forming by applications of higher flexibility dimensions

P. Groche[1,a], M. Kraft[1,b] S. Schmitt[1,c], S. Calmano[1,d], U. Lorenz[2,e], T. Ederer[2,f]

[1] Technische Universität Darmstadt, Institute for Production Engineering and Forming Machines, Petersenstraße 30, 64287 Darmstadt, Germany

[2] Technische Universität Darmstadt, Department of Mathematics, Optimization, Dolivostraße 15, 64293 Darmstadt, Germany

[a]groche@ptu.tu-darmstadt.de, [b]kraft@ptu.tu-darmstadt.de, [c]s.schmitt@ptu.tu-darmstadt.de, [d]calmano@ptu.tu-darmstadt.de, [e]lorenz@mathematik.tu-darmstadt.de, [f]ederer@mathematik.tu-darmstadt.de

Keywords: uncertainty, flexibility, metal forming, servo press

Abstract. It is widely accepted that fluctuations in market demands and product life cycles are often unpredictable. Based on these uncertainties, companies cannot calculate with constant demands. Manufacturers are also confronted with quality fluctuations in semi-finished parts that lead to various product qualities. This paper identifies the most relevant uncertainties for companies and gives answers how manufacturers can deal with these problems. It also shows recent developments in the field of flexible forming using servo press technology. Hereby the focus is set on 3D Servo Presses, providing various options for accomplishing uncertainties.

1. Introduction

Fluctuating demand scenarios for customized goods and the request for low cost - high quality products highly affect the modern global market [1]. Furthermore, the established product life cycle is shortened due to worldwide competition in the production sector [2]. In addition to that, other influence factors on the company and the manufacturing processes, such as scarcity of resources or labor availability, are commonly neglected or assumed as constants. Uncertainties caused by the influences mentioned above directly affect the production planning process as well as the production process itself. Forming companies, usually characterized by high investment costs and development effort for machines and processes, are especially subject to uncertainties caused by fluctuating demand as well as changing customer requirements. To ensure their productivity it is necessary to provide flexible manufacturing systems, to enable quick response to changing boundary conditions.

Besides uncertainties in marked demand, manufacturers are confronted with another class of uncertainties: those in quality of supply material, which deals as input for the considered production process. In forming processes, commonly used semi-finished parts are sheet metals and tubes. Quality criteria of those semi-finished products are widely standardized and should be agreed upon between supplier and manufacturer. Nevertheless, standards admit a tolerance band in which geometric and material properties can range and are therefore a challenge for the dimensioning of manufacturing processes. Complex processes often have narrow process limits in which parameters can range and offer high sensitivity for disturbing influences. Narrowing property specifications of supply material can quickly end in higher material cost and therefore minimize the manufacturer's profit. Accordingly, it is desirable for the manufacturer to use a process insensitive towards all mentioned uncertainties.

2. Uncertainties in Forming Technology

When setting up a manufacturing system, production planners are confronted with uncertainties caused by the limited validity of assumptions about the development in demand market and other conditions relevant to production, like length of product life cycles and augmented claims for low expenses and high qualities [1,3].

In case uncertainties are unavoidable, they have to be reacted to in an efficient way. Traditionally, in case of a change in production, large setup times and costs as well as downtimes are generated. Increasing flexibility of manufacturing systems and processes can enable the manufacturer to react to future circumstances properly. A flexible system is characterized by providing a wider range of possibilities to the manufacturer [1,3].

Regarding assembly and cutting machines, flexibility has been investigated intensively in the past and can be considered state of the art. This does not apply to forming machines, though. So far, forming machines are, to a large extent, predetermined in their use being either optimized for large batch productions with fixed selected forming methods or for small batch productions with predetermined specialized tool movements. Consequently, their flexibility in terms of adaptation to changing market conditions is very limited [4].

Son and Park [5] categorize flexibility into four classes:

- **Equipment flexibility**: The capability of a system to integrate new products and variants of existing products.
- **Product flexibility**: The adaptability of a production system to variances in the product mix.
- **Process flexibility**: The adaptability of the system to changes in part processing, for example caused by changes of technology.
- **Demand flexibility**: The ability of a manufacturing system to respond to changes in market demand.

Extending this fundamental approach of flexibility to forming processes and associated machines leads to the significant relevance of the number and types of driven degrees of freedom (DoF) used for the relative movement of tool and workpiece as a measure of flexibility.

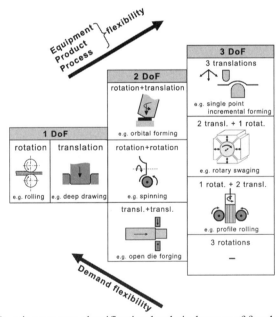

Figure 1: Forming process classification by their degrees of freedom (DoF) [6]

Figure 1 shows the relation between the number of driven DoFs and the influence on the four types of flexibility. Equipment, product, and process flexibility rise with the increasing number of DoFs whereas demand flexibility decreases especially for high volume production. This is a consequence of the increase of tool path lengths and tool-workpiece contact times leading to a reduced output.

Discussing the aspect of total flexibility, the basis for consideration is given by a movement with one driven rotational or translational DoF. For the accomplishment of the tool or workpiece movement only one drive and one control system are needed. Because of the short tool path length and contact time of tool and workpiece, the productivity and thus demand flexibility is on a high level. In contrast, equipment flexibility is small because of the shape of the predefined cross sections which is only changeable in a narrow bandwidth. Product flexibility is moderate because of long set-up times and expensive tool modifications. At last, process flexibility is poor because tools are usually dedicated to one particular process [6].

Moving one step to the right in the model (Figure 1) leads to processes with two DoFs, resulting in a higher product and process flexibility compared to single DoF processes. With the usage of an increased number of driven DoFs the tool path length and the time of tool contact are increased compared to a movement with only one DoF. Additionally, the effort to adjust two motion axes to each other is higher and requires a more complex control. These facts result in a decreased productivity of forming processes with two DoFs and therefore a reduced demand flexibility if it comes to large batch production. Several examples for this type of operation can be found in the group of incremental bulk forming processes [7].

The highest level of equipment, product and process flexibility is achieved by forming processes with three DoFs in the tool-workpiece relative motion. Using a simple tool shape and a specific tool movement allows forming products with a three-dimensional geometry. As a result, integration of new products or variants into the product mix does not require a long set-up time. On the contrary, the productivity of processes with three driven DoFs is low, leading to small demand flexibility, if large numbers of a product are required. This is due to longer tool paths and contact times between tool and workpiece as well as a more complex coordination and synchronization of the three DoFs [6].

3. Servo Presses

Since the 1990s the development and construction of servo motor driven presses is possible. This was enabled by the availability of increasingly capable torque motors.

The use of servo motors allows a process optimization by predefining motion curves to control the ram movement during forming processes. Another issue is the possibility to combine the advantages of way- and force bound presses. Thereby various forming methods have been proposed for enhancing productivity, improving product accuracy, reducing working noise, saving energy, and to extend working limit. It can be concluded that servo drive presses have capabilities to improve process conditions and productivity in metal forming.

In the following a synthetic account of the main four components of such presses is given. This includes the servo motors, the controls and sensors, the mechanics and the energy retrieving.

Servo Motors: "A torque motor can be seen as a large scaled servo motor with a hollow shaft and optimized torques." The differences between an asynchronous motor and a servo motor become obvious by considering the two characteristic torque curves and their operating window. It is apparent that the torque motor can bring its full moment even at a speed of zero. This is possible because the torque motor has no slip, which is the reason why servo presses are primarily driven by torque motors [8,9].

Controls and Sensors: In servo presses a variety of sensors can be installed that can provide information on force, position, speed, acceleration, and temperature of the defined press components. The sensors are mainly used to control the ram movement and monitor the press status. Thus, for example the current position of the ram can be given to the control. Then, a comparison of the set data and the measured data takes place. If the data differ from each other the control sends a new control signal to the servo motor(s). The conversion of the input data and controlling of the motors is a simultaneous process. This type of regulation is called a closed loop control. Therefore, it is possible to compensate ram tilting by eccentric load. This is possible due to the two separately controlled servo drives linked with the ram. The presented opportunities require a high qualification of the operating personnel [9].

Energy Retrieving: To provide the full press capacity by servo motors, a large amount of energy is taken from the electric circuit. Analyses of different operating press modes showed, that most of the energy is required during the forming and acceleration processes. It is also known, that a large amount of electrical energy is set free during the deceleration of the ram. Currently, there are two opportunities to recover the energy. It can be fed back to energy storages like capacitors or it can be stored in flywheels. This stored energy can be used for a subsequent acceleration of the press, resulting in a reduction of power consumption from the electrical network [10,11].

Mechanics: In principle, the usage of servo motors in servo presses can be divided into two categories: They can be used as high speed motors or they can be used as low speed-high torque motors. High speed motors with relatively low moments are used to transform the rotational motion into the linear motion of a press slide using a separate gear. The low speed-high torque servo motors are deployed to allow direct drive of the press eccentric or crank mechanism. This has the advantage that there is no need for belt, linkage or ball screw drives [12].

Advantages of Servo Presses: The possibility to generate a flexible ram movement such as it is known from hydraulic presses and simultaneously achieving the speeds and accuracies of mechanical presses is the great advantage of servo presses. To increase workpiece quality and output rate, an adjustment of the ram movement and velocity at constant press capacity is necessary. These opportunities are now possible by servo presses. There is also a variety of other positive characteristics such as controlling the deformation speed during forming that leads to reduced friction heat. Furthermore, the control of the dwell time at the Bottom Dead Center (BDC) and the implementation of secondary operations in the same press by slowing down or stopping the press slide anywhere [12].

The advantage of an adaptable ram motion is explained in the following, referred to Figure 2: The figure shows a comparison of the press cycle of a mechanical press with one of a servo press. Sector (1) shows the cycle time for a servo press. The aforementioned flexible programming and the closed loop control allow variable stroke length (2) and an increased deformation rate (3). Also, the accuracy can be improved (4) by dwelling the tool at BDC when another process put in simultaneously (4). Standstill at the BDC also reduces the noise and shock when the tools move apart (4). Finally, the stroke can be optimized for transfer requirements (5) [13].

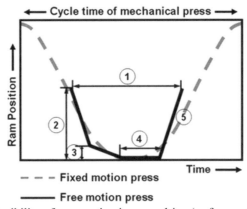

Figure 2: The flexibility of ram motion in servo drive (or free motion) presses [13]

3D Servo Press: Based on the already mentioned advantages of servo presses, a new type of servo press was developed at the Institute of Production Engineering and Forming Machines (PtU) in Darmstadt. The 3D Servo Press provides a flexible ram motion with three DoFs. The machine has a press capacity of 1 t and a max. stroke length of 40 mm. With this machine, it is possible to meet the requirements for flexibility as they were shown in Figure 3. This results in completely new manufacturing opportunities. The concept of the press could be a further step to flexible forming technology. In the following, a description of the design is given, based on Figure 3.

Due to its special layout the press can operate as a way or force bound machine. For special applications like wobbling, a combination of both operating modes is possible. The way bound mode is realized by three equal drive systems arranged star-shaped with an offset of 120°. Each of these systems consists of a servo motor and a crank mechanism. If the three drives (α_i) run in phase they are able to perform high stroke rates as already known from conventional mechanical presses. In case the three drives are controlled phase-shifted, a free translation of the tool center point (TCP) in z-direction and a rotation around the x- and y-axis (φ_1 and φ_2) can be realized. The linkage between the drive systems and the ram is realized by a connection rod and bearings which allow free rotational and translational movements. The force bound mode is realized by two spindles also driven by servo motors. The spindles are located centrally between the crank mechanisms and linked with the gear of the way bound part of the press. Through movement of the spindles, the ram is driven independently from the three drive systems in vertical direction. In addition, the spindles can also be used to vary the installation height and the stroke length of the way-bound part of the press. Examples that illustrate the flexibility of the press can be found in [14,15].

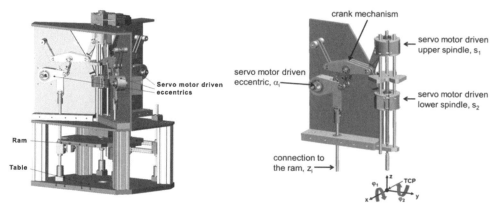

Figure 3: Mechanical structure of the 3D Servo Press [6]

The PtU is working on the implementation of a new 3D Servo Press with larger forces and a larger ram movement. This completely new developed machine will have a nominal force of 1600 kN. The height of stroke is about 300 mm. The servo motors for the eccentrics have a torque of 3500 Nm and a revolution speed of 300 $^1/_{min}$. The motors for the spindles have a torque of 6500 Nm and a revolution speed of 600 $^1/_{min}$. The table size is one square meter.

4. Variety of uncertainties in demand scenarios from the economical point of view

In forming companies, processes and process chains are optimized to maximize product output in order to enable an economic production. Boundary conditions and requirements are generated by customer demands and estimations. The forming process, the whole process chain and the company itself are subject to various influences which cause effects and necessitate activities. Activities and effects can be distinguished according to the location of their appearance. Internal effects emerge inside the companies' or the process' boundaries. The environment influences the company with its production processes by external effects. While activities inside the company can be influenced in a

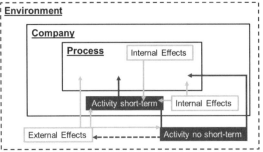
Figure 4: Origin of effects and activity [16]

short time period, activities outside the company's direct sphere of influence are institutional long term decisions. Effects are process relevant results of changes, whereas activities are responses to occurring change [16]. In Figure 4 the origins of effects and activities are displayed.

The occurrence of uncertainty is hereby specified as the cause of change. To analyze the whole spectrum of uncertainties and the concurrent effects, it is necessary to define terms in order to develop a classification. For this reason four cases of change are identified and their interdependencies are displayed in Figure 5. On the process level for example, an external technological effect may result from the availability of a specific technology. Wear of tools or heat development might cause deviations which will cause varying process outcomes although the input parameters remain stable. This is defined as a process internal effect. Sociological and economic effects occur within as well as outside the company whereas natural effects cannot originate from inside the company [16].

Cause of change / Result		Technological	Natural	Sociological	Economical
Effect		Process-Internal Exceeding of tolerances due to wear	Scarcity of resources	Company-Internal Shift in age structure	Company-Internal Temporary fluct. of reserve fund
		External Varying quality semi-finished products		External Alternating labor availability	External Changes in demand / supply
Activity	in-company influence-ability	Directive of inspection	Selection of alternative material	Continuing education	Development of product variants
	institutional influence-ability	Standardised certification	Introduction of environ. laws	Changes in education system	Collection of duties

Figure 5: Areas of uncertainty with examples [16]

For this purpose flexible forming systems, such as the 3D Servo Press mentioned in the chapters before, are developed. This production system provides the possibility of using multiple degrees of freedom to produce a flexible number of pieces. The selective shutdown of various degrees of freedom also enables the press to generate a high output. With these possibilities a wide spectrum of demand scenarios can be achieved. The uncertainty of fluctuations in demand can be controlled by the application of multiple technologies on the 3D Servo Press. Therefore, solely a change of the tool is required. A conventional machine would incur higher costs caused by additional investments in machinery. In the following, an example including a product as well as various machines and production processes is analyzed to compare conventional machines with the flexible 3D Servo Press in case of fluctuating demand scenarios and uncertain demand forecasts.

Three forming processes are developed to produce the plain bearing shown in Figure 6. Orbital forming is hereby applied for small lot sizes. Therefore, a tube is used as a semi-finished part to generate the bearing by a three-dimensional tool movement before it is ejected from the die. The production process is characterized by the lowest investment for tool and machine as well as minor set-up costs. However, the semi-finished product causes high costs and the cycle time is long due to the incremental tool movement. Direct tubular impact extrusion is used for medium lot sizes and requires also a tube as a semi-finished part. Consequently, the costs for the material are as high as mentioned before. The cycle time is decreased and the investment for machine and tool is higher compared to orbital forming. In case of mass production, backward can extrusion with punching and bulging is commonly applied due to its lower production times. The semi-finished part for the third option is solid metal, which decreases the material costs. Among the three production alternatives the investment and set-up costs are the highest as a result of the complexity of tool and machine [17].

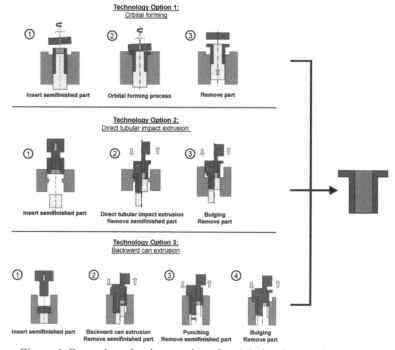

Figure 6: Example technology options for plain bearing production [17]

All three production options require a specific single technology machine and corresponding tools. A three-dimensional ram movement is essential for the orbital forming process whereas backward can extrusion requires a higher stroke length compared to direct tabular impact extrusion. All of the conventional machines are highly specialized on the specific process and on a constrained demand range. On a flexible multiple technology machine, such as the 3D Servo Press, all of the shown technologies can be implemented and the range of an economic production lot size is extended. Various demand scenarios with and without uncertainties are analyzed in an optimization model to compare the 3D Servo Press to the conventional machines [17].

The model analyses the internal rate of return (IRR) of the investment in machines, tools and materials to find the most economically advantageous technology set. A grid with deterministic constant values in a defined number of time periods is generated to analyze the demand. To implement uncertainties in the demand for every time period the lot size is assumed as a probability distribution represented by discrete values. An example for such a deterministic demand grid (DDG) and a stochastic demand grid (SDG) is shown in Figure 7 [17].

Figure 7: Demand grid without (left) and with (right) uncertainties [17]

The model calculates the quantity of cases in which a conventional machine set or a flexible production system is chosen. Depending on the sales price and the storage costs it becomes apparent in Figure 8 that the number of selected flexible machines increases if uncertainty is considered [17].

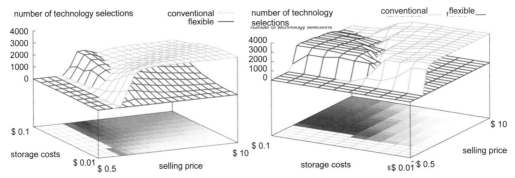

Figure 8: Number of conventional and flexible machine selection
without (left) and with (right) uncertainties [17]

The spread between the best and the worst possible IRR is divided by the expected value for the IRR to analyze the effect in worst case scenarios. It emerges that the relative deviation of the IRR is remarkably lower in cases in which the flexible machine is employed (Figure 9). Hence, the employment of flexible technologies such as the 3D Servo Press increases a company's protection against losses caused by fluctuating demands [17].

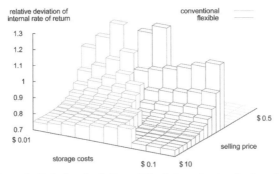

Figure 9: Relative deviation of the internal rate of return [17]

5. Adaptive and closed-loop control of forming processes

As stated before, many manufacturing processes are widely sensitive to fluctuation in material properties. The method to reduce sensitivity is to increase the flexibility of forming processes without a significant decrease in productivity.

For various established forming processes, different approaches of facing uncertainties in process-relevant properties of the semi-finished part, which is the input for the manufacturing process, can be defined. They all have in common, that information about the actual condition of the machine, tool or workpiece is fed back to the process control and utilized to enhance the process and achieve an improvement in product quality.

Due to the variety of forming processes, not all approaches are applicable for all processes, but can be combined and cascaded, leading to a strong mutual dependence.

Approach I: Incremental closed-loop process control. A process is conducted with a standard set-up; after each stroke, the quality of the final product is measured and assessed. From this analysis, a process modification is deviated, aiming for a quality improvement of subsequently conducted processes.

As an example the process of blanking may serve. Depending on the sheet thickness of input material, process forces and velocity curves have to be adapted. By measuring the quality of the output, the process can be held in its functional range as long as changes in input properties are gradual and do not immediately cause the process to collapse.

For realization of this approach, a sensor is needed to quantify the output quality. This information is processed by the open-loop processor, adapting the set-point curves for the velocity or force.

This approach does not require a deep knowledge of the correlation between input properties, process values, and output properties. Disadvantageously, an adaptation to fluctuation in input properties can only be made if they have already had an influence on the output quality.

Approach II: Input Property adaptation. If the comprehension of the process values is existent, at least by trend, the process can be adapted to the actual input properties before it is conducted. In case of the above mentioned blanking process, this approach includes measuring the sheet thickness of the input material beforehand and adapting the velocity or force curve of the following stroke based on the known correlation functions.

Another example for this approach, making use of the multiple degrees of freedom provided by the 3D Servo Press, is deep drawing of non-uniform thickness distributed blanks. In conventional presses, die and blank holder are held parallel, leading to a non-uniform distribution of contact pressure on the sheet. This can cause an excessive contact pressure in areas of higher sheet thickness and wrinkle forming in areas of lower sheet thickness. To prevent this, the actual thickness of the blank sheet is measured at various points and supplied to the control. As a reaction, the press control presets an appropriate inclination of the platform to which the die or blank holder are attached and can therefore take influence on the proper conduction of the deep drawing process. This can lead to a significant decrease in wrinkle forming.

Approach III: Internal feedback based position and velocity adjustment. This third approach is based on and can be described with the classic closed-loop feedback control theory. It is necessary to adjust tool path or velocity to the set-point values provided by the open-loop processor and make the system non-sensitive to disturbance values. The system to be controlled is the tool motion, being driven by the drive chain of the press and disturbed by process forces. The actual position and orientation of the platform can be measured by linear sensors and fed back to the closed-loop processor, which generates an actuating variable. This correcting value is superposed with the set-point motion values from the open-loop processor to adjust the tool position to the desired value. Figure 10 shows the material and signal flow as well as all described elements involved in all three approaches.

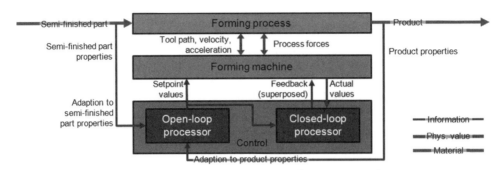

Figure 10: Multiple approaches for press control

In summary, the three mentioned approaches need to be combined to create a holistic production system which is widely insensitive to uncertainties in input material properties, acting as disturbances on output quality and process stability.

Conclusion

As shown in this paper, manufacturers are confronted with a variety of uncertainties. The demonstrated approach is to increase flexibility of manufacturing processes without a significant reduction of productivity. New developments in servo press technology, especially the availability of 3D Servo Presses, enable forming companies to be prepared for uncertainties in demand markets and semi-finished part properties. Innovative control strategies lead to an adaptive reaction of the forming process to the actual state of fluctuating input parameters and resolve technological challenges thereby. This current approach appears to be auspicious especially for forming technology and will be examined in further research.

Acknowledgements

The investigations presented in this paper were carried out within the research project B2: "Forming – Production Families with Uniform Quality" of the Collaborative Research Centre 805 "Control of Uncertainties in Load-Bearing Mechanical Engineering Structures" (CRC 805). The authors thank the German Research Foundation (DFG) for founding and supporting the CRC 805.

References

[1] G. Chryssolouris: Manufacturing Systems – Theory and Practice, Second Edition, Springer, New York, 2006

[2] H.-P., Wiendahl, et al.: Changeable Manufacturing – Classification, Design and Operation, CIRP Annals, Vol. 56/2, 2007

[3] D. Gerwin, Manufacturing Flexibility: A Strategic Perspective, Management Science 39(4):395–410, 1993

[4] HA. ElMaraghy: Flexible and Reconfigurable Manufacturing Systems Paradigms, International Journal of Flexible Manufacturing Systems 17(4):261–276, 2006

[5] YK. Son, CS. Park: Economic Measure of Productivity, Quality and Flexibility in Advanced Manufacturing Systems. Journal of Manufacturing Systems 6(3):193–206, 1987

[6] P. Groche et al.: Increased total flexibility by 3D Servo Presses, CIRP Annals - Manufacturing Technology, Volume 59, Issue 1, 2010

[7] P. Groche, D. Fritsche, EA. Tekkaya, J. Allwood, G. Hirt, R. Neugebauer: Incremental Bulk Metal Forming. Annals of the CIRP 56(2):635–656, 2007

[8] D. Schröder: Elektrische Antriebe 1. Grundlagen (engl. Electrical Drives 1. Fundamentals), Springer Verlag, Berlin, Heidelberg, New York, Tokyo, 1994

[9] T. Gräbener, P. Groche, M. Kraft: Recent Developments in Servopress Technology and their Applications in Cold Forging Processes, In: 43rd ICFG Plenary Meeting 2010 September 12th to 15th, Darmstadt, Germany

[10] J. Roske: Cold Forging with Servo Presses, Schuler Pressen GmbH & Co. KG, Göppingen, 2010

[11] A. Osborn, S. Paul: Servo Press Technology: Drive Design and Performance, Metal Forming Magazine, Issue 8, 2008

[12] D. Boerger: Servo Technology Meets Mechanical Presses, Stamping Journal, Nov/Dec. 2003, p.32

[13] T. Altan: Servo-Drive Presses – Recent Developments, 10. Umformtechnisches Kolloquium Darmstadt, Verlag Meisenbach Bamberg, 2009

[14] P. Groche, M. Scheitza: Three degrees of freedom servo-press – Conceptual design and processes for flexible forming, wt-Online, Volume 97, 2007

[15] P. Groche, M. Scheitza: Erweiterte Fertigungsmöglichkeiten durch eine 3D-Servopresse (engl. Advanced Production Opportunities by a 3D Servo Press), Umformtechnisches Kolloquium Darmstadt, Meisenberg Bambach, 2009

[16] S. O. Schmitt, J. Avemann, P. Groche: Development of manufacturing process chains considering uncertainty, CARV Conference Proceedings "Enabling Manufacturing Competitiveness and Economic Sustainability", Springer, 2011

[17] J. Avemann, S. O. Schmitt, T. Ederer, U. Lorenz, P. Groche: Analysis of Market Demand Parameters for the Evaluation of Flexibility in Forming Technology, CARV Conference Proceedings "Enabling Manufacturing Competitiveness and Economic Sustainability", Springer, 2011

Early stage geometrical deviation optimization – an automotive example for sheet metal parts

Bohn, Martin[1, a]; Steinle, Philipp[1, b]; Wuttke, Fabian[1, c]

[1] Daimler AG, Benzstrasse 1, 71063 Sindelfingen, Germany

[a] martin.bohn@daimler.com

[b] philipp.steinle@daimler.com

[c] fabian.wuttke@daimler.com

Keywords: variation prediction, tolerance management, sheet metal, deep-drawing, automotive industry

Abstract. The recent emphasis on car styling caused tolerance requirements for sheet metal parts in the automotive industry to increase. In addition, the new materials (steel and alloy) have a different deviation behavior around the springback after the deep-drawing process. In the early design phases a product can be optimized to fulfill tight tolerance specifications. This work shows the simulation background and the resulting optimization methodology.

Introduction

There is a trend in the automotive industry towards lightweight concepts [1]. These concepts contain new high tech steels and alloys. To be successful in the automotive industry, it is necessary to ramp up fast with high quality [2]. To achieve this, the sheet metal parts have to be quickly within a tight tolerance range. Deep-drawing simulation tools are widely used to predict the springback effect. This effect is most critical for high strength steel materials [3].

In general, the springback is compensated by over-bending the parts [4], [5]. There is much research on robust optimization in the metal forming process [10], [11], [12]. The accuracy of the spring back for complex automotive parts with commercial tools, like Autoform Sigma, is around ±1 mm compared to measured deep-drawn parts. The reason for this is the uncertainty of the input parameters like thickness, friction properties and material properties. Therefore, hardware optimization loops are still necessary. The springback effect itself is a mean value shift. The deviation is the variation around this value. Today there are no commercial tools available to predict this precise, otherwise the springback would be calculated much more precise. Actual material models don't focus on deviation / tolerance [6], [13].

For validation of the deviation hardware is needed. The alignment between development process and hardware is shown in Fig. 1.

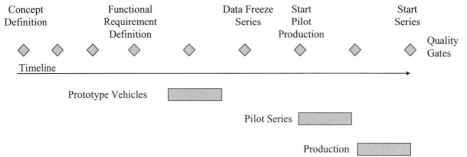

Fig. 1 Hardware phases

During the prototype phase not more than 5 parts are deep-drawn without tooling optimization. To estimate the standard deviation of a geometrical element a certain amount of samples is needed. The correlation between the amount of samples and accuracy (confidence interval) of deviation prediction is described in [7], see equation (1).

$$I(x_1 \ldots x_n) = \left[\sqrt{\frac{(n-1) \cdot s_n^2}{\chi^2_{n,\frac{1-\alpha}{2}}}}, \sqrt{\frac{(n-1) \cdot s_n^2}{\chi^2_{n,\frac{\alpha}{2}}}} \right] \qquad (1)$$

I = confidence interval
n = amount of samples
α = probability
χ = quantile of chi - square distribution
s = standard deviation

A typical example for the confidence interval based on a probability of $\alpha=5\%$ and a standard deviation $s = 0,167$ (equals a tolerance of $\pm 0,5$ mm) is shown in figure 2.

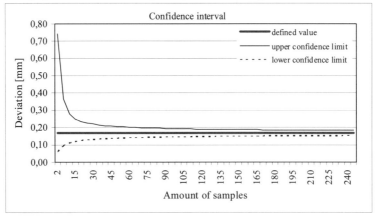

Fig. 2 Confidence interval

This shows clearly, that it is necessary to have a big amount of samples ($n > 100$) for any relevant estimation of the deviation.

The critical case is, when the result of the measurement shows a lower deviation than the reality. Therefore, the lower confidence limit is the critical line. The next figure shows the possible percentage, how much bigger the real deviation can be compared to the measured deviation.

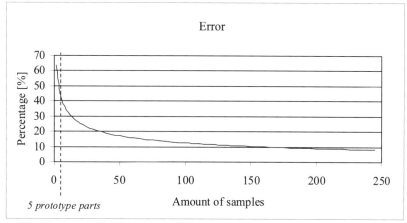

Fig. 3 Difference between real deviation and measured deviation

It is obvious, that a small amount of five samples during the prototype phase will cause severe errors.

In the early development stage, here the prototype stage, the product design can easily be modified. In a later phase the modification will be much more expensive, due to modifications in the series tooling [2]. This leads to the conflict "ease of modification" versus "uncertainty".

Therefore, a new approach is needed to optimize the geometry in early stages, even prior any hardware phase. The target here is the reduction of the deviation itself and not the reduction of the springback effect (mean value shift), because the mean value shift can be eliminated in hardware optimization loops. This is shown in the following figure.

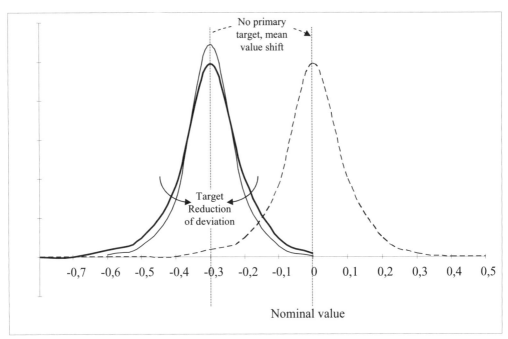

Fig. 4 Optimization target

This target differs from the optimization target in recent research work and from the optimization target of the simulation software. In addition, the parts will be manufactured from different suppliers. Therefore, the benefit from tools like Autoform Sigma is limited due to unknown process parameters across different, maybe even unknown suppliers. Hence, the mathematical aspect of the deviation prediction has to be on a higher abstraction level.

Approach

One approach to calculate the standard deviation of sheet metal parts is based on the stiffness of the parts [8]. Within the FE-model, the part is fixed at the location points and a reference force (e.g. 10 N) is applied at the measurement point. The deflection of this point (μ) is the result of the applied force. This process has to be repeated for all measurement points. The standard deviation for one measurement point (Eq. 2) consists of two parts. The first one is the standard deviation ($S_{reference\_system}$) caused by the repeatability, if you put the identical part many times in the fixture. The second part is the deviation of the part itself. The material specific values K_i have to be determined with real measurements on different parts but from the same material. Therefore, the material specific values K_i are the same for all measurement points. For practical reasons the sum can end at $i=2$.

$$S = \sqrt{S_{reference\_system}^2 + \left[\sum_{i=0}^{\infty}(K_i \cdot \mu^i)\right]^2} \qquad (2)$$

The equation leads to the conclusions:
- The reference system has to be robust.
- Bigger stiffness results in a lower standard deviation.

The analysis of the springback effect leads to similar results in related works [9]. If a high stiffness is good for precise parts, the optimization strategy in early phases should be, to design the part in tolerance critical areas as rigid as possible.

Verification and enhanced optimization process

The verification is done using the deep-drawn part of a door inner panel.

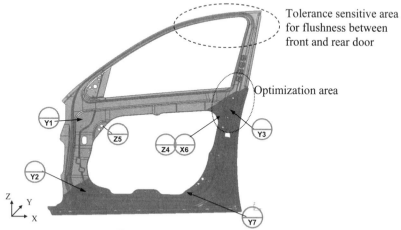

Fig. 5 Deep-drawn door inner panel

The upper rear area of the front door is very tolerance sensitive for the flushness between front and rear door, because of the high visibility before opening the door. In the early design phase prior any hardware the design of the sheet metal door inner panel is optimized to maximum stiffness.

In the prototype phase the mean value shift can be measured (amount of samples: 5) but there is no chance to predict, if the deviation will meet the tolerance requirements.

If the measurements during the pre-series show too much deviation and the quality of the part cannot be improved, there are two possible options for action. First, tolerance requirements could be widened. If that's not possible, the stiffening optimization strategy has to be withdrawn, because in the following body in white process the high stiffness disallows the bending process to achieve the target tolerance specification. Therefore, an enhanced optimization process is necessary.

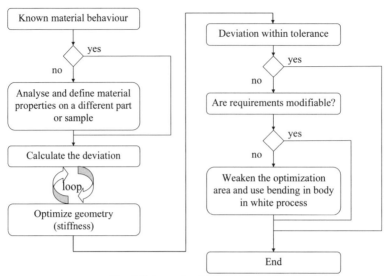

Fig. 6 Enhanced optimization process

As proved, it is necessary calculating the deviation. To do this, the material properties have to be known. After the optimization process either the goals are met or the part has to be weakened to bend it afterwards to meet the tolerance specifications. This could be done either on the part itself or during the assembly process.

Conclusion

In the early design stage, the design of a sheet metal part can easily be modified, but the verification, if the part is within tolerances, can't be done with hardware. Furthermore, there is no general optimization strategy to reach tight tolerance requirements for parts out of new materials. The sole optimization against springback may lead to a suboptimal design of the part. The only way to decide which optimization strategy will work, requires either a much bigger hardware amount of samples in the prototype phase or the analysis of the material properties on different parts and the calculation of the deviation. The required process is shown and verified.

References

[1] M. Kleiner, M. Geiger, A. Klaus, in *Manufacturing of Lightweight Components by Metal Forming*, ICED at Paris, (2007).

[2] K. Ehrlenspiel, A. Kiewert, U. Lindemann in *Kostengünstig Entwickeln und Konstruieren (engl. Cost saving design and development)*, VDI Buch, (2007).

[3] J. Danzberg in *Effizienzsteigerung durch innovative Lösungen in der Methodenplanung und im Werkzeugbau (engl: Efficiency increase trough innovative solutions in tooling design)*, SWISSMEM & IVP, ETH Zürich, (2007).

[4] K. Roll in: *Simulation der Blechumformung – neue Anforderungen und Tendenzen (engl: Deep- drawing simulation – new requirements and tendencies)*, Tagungsband 12. Dresdner Werkzeugmaschinen-Fachseminar Simulation von Umformprozessen unter Einbeziehung der Maschinen und Werkzeugeinflüsse, Dresden, (2007).

[5] M. Schroeder in: *Weitere Potenziale der Topologieoptimierung zur Rückfederungskompensation von Blechformteilen (engl: Further potentials of topology optimization for spring-back compensation of deep-drawn parts)*, Tagungsband 12. Dresdner Werkzeugmaschinen-Fachseminar Simulation von Umformprozessen unter Einbeziehung der Maschinen und Werkzeugeinflüsse, Dresden, (2007).

[6] A. Krasovskyy in *Verbesserte Vorhersage der Rückfederung bei der Blechumformung durch weiterentwickelte Werkstoffmodelle (engl: Improved springback forecast of deep-drawn parts trough improved material models)*, Dissertation Karlsruhe, (2005).

[7] E. Dietrich; A. Schulze in *Statistische Verfahren zur Maschinen- und Prozeßqualifikation (engl: Statistical methods for machine and process qualification)*, Carl Hanser Verlag München, Wien, ISBN 3-446-18780-4, (1996)

[8] M. Bohn in *Toleranzmanagement im Entwicklungsprozess (engl: Tolerance management in the development process)*, Dissertation Karlsruhe, (1998)

[9] K. Roll, T. Lemke, K. Wiegand in: *Simulationsgestützte Kompensation der Rückfederung (engl: Simulation based springback compensation)*, Tagungsband 3. LS-DYNA Anwenderforum in Bamberg, (2004).

[10] J. H. Wiebenga in *Optimization under uncertainty of metal forming process – an overview*, M2i, (2010)

[11] M. H. A Bonte, A. H. van Boogard, B.D. Carleer in *Optimising towards robust metal forming processes*, Tweente, (2006)

[12] M. H. A Bonte, A. H. van Boogard, J. Huetink in *Deterministic and robust optimisation strategies for the metal forming processes*, Forming Technology Forum 2007, ETH Zuerich, (2007)

[13] R. ter Wijlen in *Optimisation of a deep drawing processwith experimental validation, Applied to an automotive deep drawing process of a B-pillar*, Graduation assignment University of Twente, (2007)

Methods for the control of uncertainty in multilevel process chains using the example of drilling/reaming

Michael Haydn[1,a], Thomas Hauer[1,b] and Eberhard Abele[1,c]

[1]Institute of Production Management, Technology and Machine Tools (PTW),
Technische Universität Darmstadt, Petersenstraße 30, 64287 Darmstadt, Germany

[a]Haydn@ptw.tu-darmstadt.de, [b]Hauer@ptw.tu-darmstadt.de, [c]Abele@ptw.tu-darmstadt.de

Keywords: uncertainty, reaming, process simulation

Abstract: Uncertainty during production processes has an important influence on the product quality as well as production costs. For multilevel process chains with serially connected processes, additional uncertainty can be caused by the previous step. The manufacturing of precision holes by drilling and reaming is an important multilevel process chain. The interactions between machine, tool and pre-drilled hole cause process errors during the quality determinant final reaming process. In this paper, a systematic approach for the identification and control of uncertainty during the reaming process is presented. Thus, the influence of key aspects like skewness of pre-drilled hole or the influences of material strength gradients are analyzed. Further, simulation models for the consideration of these uncertainties are presented.

Introduction

Products of mechanical engineering are exposed to uncertainties which significantly influence the product characteristics during the product lifecycle. These uncertainties, e.g. wrong assumptions due to missing information in development, material disturbances during production or wrong load spectrum in the usage, can have disastrous consequences. Thus, nowadays safety factors, for development or usage, and exorbitant tolerances, for production, are the reaction on uncertainties.

The aim of the Collaborative Research Center (CRC) 805, sponsored by the German Research Foundation (DFG), is the analysis and control of uncertainty in all processes of the lifecycle. The first phase concentrates on the development, production and usage processes for load carrying systems in mechanical engineering. The key for the uncertainty control are the methods and technologies which are being developed and verified within the CRC 805.

Drilling is an important production technique inside production processes, more precisely the machining technology. For example, drilling operations for the machining of a cylinder head in automobile production, consume nearly 89% of the production time [1]. Thereby, the production of precision holes takes about 42% of the overall production time, the highest for any operation. For this, a multilevel process chain comprising drilling and reaming processes is necessary. Thus, due to the prior drilling process, additional uncertainties, e.g. an incorrect predrilled hole, can be expected for the reaming process. This presents a problem because the manufacturing of precision holes mostly takes place at the end of the value chain. Therefore, uncertainties and the resulting failures during drilling and reaming have not only an important influence on the resulting hole quality, but can also lead to high costs.

Uncertainty in drilling/reaming is the focus of the research work at the Institute of Production Management, Technology and Machine Tools (PTW), which takes place in the CRC 805. The main stress is set on the reaming process, because it is the quality determinant key process and has to deal with increased uncertainty. This is caused by the upstream pre-drilling process and its process errors, which could have a significant influence on the reaming process. Thus, in the first step, machining tests are conducted and tool benchmarks are performed for the systematical identification of uncertainties. Based on the test results, process models can be made and simulations performed to analyze uncertainty effects. With these simulations the development of different tool, machine and process based methods for uncertainty control is possible. In the following some of these methods will be described in detail.

State of research

Generally the uncertainties during production processes can be sorted into categories derived from the cause and effect diagram according to Ishikawa [2]. These categories, often called 5Ms, are milieu, man, method, machine and material. As shown in Fig. 1, all these factors influence the product quality.

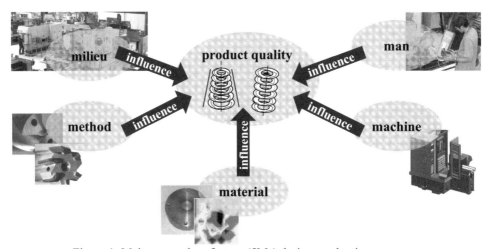

Figure 1: Main uncertainty factors (5Ms) during production processes

All uncertainties resulting from the environment, for instance a circadian temperature gradient in the plant, separately excited vibrations caused by other machines or air humidity effects, are listed within the category 'milieu'. The category 'man' includes influences caused by the human element. These can be programming errors or clamping errors for manual charging made by the machine operator. Uncertainties arising from the production technique, e.g drilling, reaming or milling, influence of process parameters or different tool geometries, are summarized in the category 'method'. Uncertainties generated by the machine are for example positioning accuracy and repeatability of the machine axes and internal heat or vibration sources. In the category 'material' all uncertainties resulting from material properties are listed. These can be material inhomogeneity, casting defects, e.g. blowholes, and strength gradients. The main research at the PTW focuses on uncertainties caused by the categories method, machine and material.

As mentioned above for the production of precision holes, a multilevel process chain consisting of drilling and reaming is required. Thus, additional uncertainties follow in the downstream operation – the reaming process. In consideration of the state of research, the main uncertainty factors for reaming can be divided into the parts misalignment and runout error. In addition, a third uncertainty factor has to be listed, the skewness of the pre-drilled hole (Fig. 2). This aspect has not been focus of research projects yet.

The reason for misalignment and runout error is the interaction of the machining process with the machine and tool uncertainties. Misalignment is a displacement of the reamer axis and the pre-drilled hole axis, which could be caused by an insufficient positioning accuracy of the machine tool. A runout error could be generated by an eccentric clamping of the tool in the tool holder or the motor spindle. Furthermore, other causes of fault could be a geometric error of the tool or a runout error of the spindle bearings. These two main uncertainty factors have been considered in various research projects in the past [1,3,4].

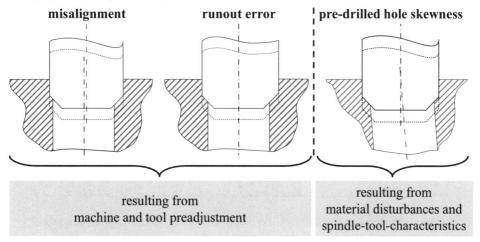

Figure 2: Main uncertainty factors and sources during the reaming process

The third uncertainty factor for the reaming process, the skewness error of the pre-drilled hole, is caused by an interaction between drilling process, work piece material and tool stiffness. Material inhomogeneities or strength gradients in the material lead to unbalanced loads at the drilling tool. This causes a displacement of the drilling tool leading to a skewness error in the pre-drilled hole. As mentioned above, the skewness error and its effects on the reaming process are actually not part of the state of research. However, this skewness could be a major uncertainty during the milling process and therefore, an important research focus at the PTW in the CRC 805.

Benchmark test of tools used in manufacturing precision holes

The first step for uncertainty control is the identification of the appearing uncertainty in machining processes. Therefore, it is important to know the capability of the production technologies, the process chains and the used tools. The determination of this aspect is realized by machining tests for the manufacturing of a precision hole with diameter 14H7 at the PTW.

In these spot tests, two different drilling tool types (solid carbide twist drill and replaceable head drill) and a multiblade reamer are used. The material for the machining tests is heat-treated steel 42CrMo4V (1.7225). The process parameters for the tests are taken from the manufacturer's instructions. For each measurement series, tools from different batches are taken to reduce the influence of the batch. Furthermore, a test is conducted during which a group of holes at random positions in the workpiece is drilled. This reduces the influence of workpiece material.

For the assessment of the tool and process capability, the resulting hole quality is estimated by geometric product specifications (diameter, roundness, cylindricity, radial deviation) from DIN EN ISO 1101 [5]. The analysis and comparison of the test results (Table 1) shows the capability of the different tool concepts.

An inspection of the achieved diameters shows the best results for the multiblade reamer. All reamed holes are within an ISO tolerance grade IT 7 and fulfill the requirements concerning the diameter. With a tolerance grade of IT 9 to 10, the replace head drill provides the worst result. Generally the best results are achieved by a multilevel process chain consisting of drilling and reaming processes. This becomes apparent by the significant improvements in diameter, roundness and cylindricity after the reaming process

Table 1: Comparison of the hole quality produced by different tool types

tool concept / criterion	solid carbide twist drill	replaceable head drill	solid carbide twist drill + multi blade reamer
diameter (IT)	IT 9	IT 9 / IT 10	IT 7
roundness [µm]	10 – 15	13 – 19	3.5 – 5
cylindricity [µm]	28 – 52	50 – 64	9 – 11

The diagram of the radial deviation of the used drill concepts (Fig. 3) shows the results of the spot tests. The skewness of the pre-drilled hole can be up to 0.2 mm depending on the drilling tool concept used. This could be a major uncertainty during the following reaming process.

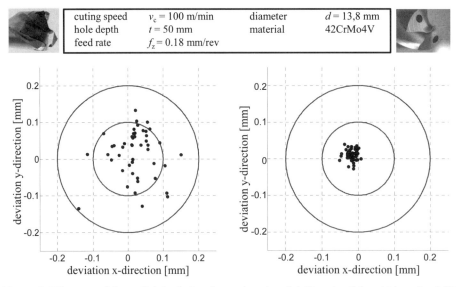

cuting speed $v_c = 100$ m/min   diameter $d = 13{,}8$ mm
hole depth $t = 50$ mm   material 42CrMo4V
feed rate $f_z = 0.18$ mm/rev

Figure 3: Diagram of the radial deviation for replace head drill and solid carbide twist drill

To sum it up, the machining tests show different uncertainty potential for different tool concepts. Thus, in the first step, the uncertainties during the production of a precision hole can be reduced by the right tool selection. Furthermore, the multilevel process chain (drilling/reaming) delivers the best results. However, for this case, the skewness of the pre-drilled hole (up to 0.2 mm) could be an important additional uncertainty during the reaming process. Therefore, it should be focused on in further research. In addition, the tests show that the material strength gradient is also a key uncertainty, as it leads to large variations in the resulting hole quality. Presently, the material strength gradient is neglected in the models for the simulation of the reaming process. Thus, this effect has been focus in further research, to analyze its influence in detail and increase the model accuracy, e.g. for force models.

Influence of the pre-drilled hole on the reaming process

According to standard literature, reaming is defined as a finishing technology for the improvement of the hole quality, which has no influence on position or form errors [6,7]. As a result the reamer follows the pre-drilled hole [4,8]. At first sight, this statement seems to agree with the results of the tool benchmark. However, to explore the influence of the skewness of the pre-drilled hole, further research is conducted at the PTW. Machining test are performed with a hole diameter of 14H7 and a variable pre-drilled hole skewness up to 0.2 mm. The used tool types are solid carbide twist drills and multiblade reamers with different tool geometries (lead angle, flank width, backtaper ratio). The work piece material is again 42CrMo4V (1.7225).

Before a discussion of the test results, the possible mechanism which leads to the deviation of the reaming tool will be presented. The tool deviation of the reamer results from the interaction between the chip cross-section and the resulting cutting forces. For an ideal reaming process, without the uncertainty of the skewness, the chip cross-section at each blade is equal. Thus, the resulting radial force F_{rad}, e.g. for a multiblade reamer with six blades, is 0,

$$F_{rad} = \sum_{i=1}^{6} F_{p,i} + \sum_{i=1}^{6} F_{c,i} = 0. \tag{1}$$

For a skew pre-drilled hole the chip cross-section is not equal at all cutting edges (Fig. 4). This leads to varying passive forces at the different blades. Fig. 4 shows the cutting forces for opposed blades with maximum ($F_{p,1}$, $F_{c,1}$) and minimum ($F_{p,2}$, $F_{c,2}$) width of cut. In this case, the resulting radial force $F_{rad} > 0$ N leads to a tool deviation in the direction of the second blade at the beginning of the reaming process. As a result, in theory, the reamer follows the skewness of the pre-drilled hole.

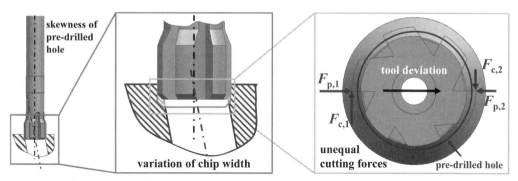

Figure 4: Possible reason for the reamer deviation by pre-drilled hole skewness

The results of the machining tests show that the theoretical consideration and the conventional wisdom are not applicable in all cases [9]. For all used tools and skewness values, the reamer does not follow the pre-drilled hole for the chosen diameter 14H7 and L/D-ratio of 8. Even in the worst case scenario, when an aircut appears, the deviation of the reamed hole is only 34 µm for a pre-drilled hole with a skewness of 200 µm, thus less than a quarter of the expected value (Fig. 5 left). The test results also show the important influence of the measuring strategy (Fig. 5 right). The first strategy captures parts of the pre-drilled hole for an appearing aircut. The effect is a radial deviation of 48 µm at a hole depth 48 mm. Thus, a modified measuring strategy is chosen which considers the area containing the reamed hole only. This leads to a radial deviation of 34 µm at 48 mm which is an improvement of about 30%.

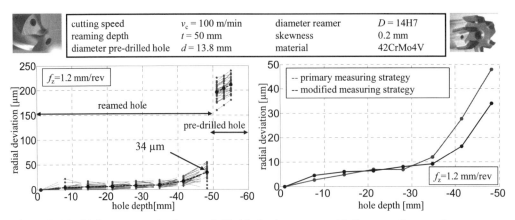

Figure 5: Machining test results for pre-drilled hole skewness and influence of measuring strategy

As mentioned above, the conducted research shows that the conventional guess concerning the uncertainty skewness has to be limited. The results of the machining test, for the first time, show that in special cases the reaming tool is robust against the pre-drilled hole skewness. Thus, besides the resulting process forces, the stiffness of the spindle-toolholder-tool-system, especially the tool and toolholder stiffness, has to be considered. The influence of the diameter and the L/D-ratio on the tool stiffness and tool deviation should be inspected and a limit should be determined. Furthermore, the flank width also has an important influence on the tool deviation. The theoretical considerations and machining tests also show the need for a simulation model which considers the chip cross-section, resulting cutting forces and the flank influence for the control of uncertainties during production processes.

Simulation of the chip cross-section

An important step in the control of uncertainties during the production processes is the knowledge about their effects. The analysis can be carried out in two different ways. The first is to carry out cutting experiments which give a good overview about the uncertainties which could appear and their effects. But as a disadvantage they are time and cost intensive. The second method is the usage of simulation models which offer the possibility of analyzing different uncertainty effects in a cost effective way. The quality of the results depends on the assumptions which were made before. Thus, it is also necessary to verify simulation results by machining tests. The optimal method is to combine simulations, a cost effective analysis of different influence coefficients, with machining tests, for the verification of the simulation results and the improvement of the simulation quality.

As mentioned above, the cutting forces and the chip cross-sections are important for the simulation of uncertainties during the production processes. Uncertainties during the reaming process, e.g. misalignment, runout error and the skewness of the pre-drilled hole, have a direct influence on the chip cross-section at each cutting edge. Thus, the first step for the simulation of uncertainty is the creation of a chip cross-section model which gives the possibility to consider appearing process errors during the machining process.

The chip cross-section

$$A = b \cdot h = f_z \cdot a_p, \qquad (2)$$

is calculated by multiplying the chipping thickness h with the width of cut b. Alternatively, the feed per tooth f_z and the depth of cut a_p can be used. For the detailed calculation of the chipping thickness

$$h = f_z \cdot \sin \chi \qquad (3)$$

the entering angle χ is necessary. For an ideal reaming process the width of cut

$$b = \frac{a_p}{\sin \chi} = \frac{D-d}{2 \cdot \sin \chi}, \qquad (4)$$

determined by the diameters of the pre-drilled hole d and the reaming tool D, is constant and equal at every cutting edge (Fig. 6).

The result of uncertainties during the reaming process is a varying chip cross-section which is caused by, for instance a varying width of cut b. The reason for this is a changing depth of cut a_p at every cutting edge. For example, the skewness of a pre-drilled hole leads to a minimum b_{min} and maximum b_{max} width of cut at the cutting edges during one revolution (Fig.7 left).

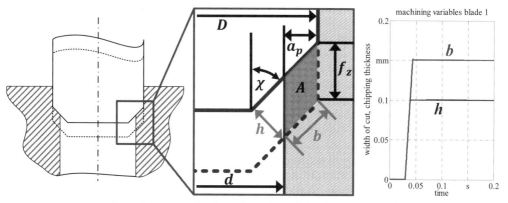

Figure 6: Factors for the calculation of the chip cross-section

The open loop simulation of the machining variables during the reaming process in combination with the uncertainty of the pre-drilled hole (Fig. 7 right) clarifies the effect of the fluctuating width of cut. With increasing simulation time the deviation of the width of cut during a revolution rises. The thickness h, which is dependent on the feed per tooth f_z, is constant.

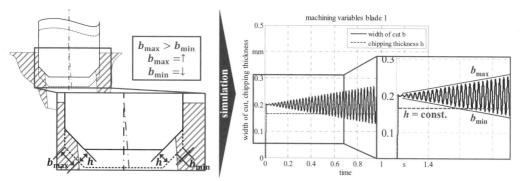

Figure 7: Open loop simulation of the chip cross-section for skewness

The chip cross-section simulation, developed at the PTW, has the ability to take into account different uncertainties like the described skewness of the pre-drilled hole. Thus, it is possible to test the influence of different machining parameters on the chip cross-section and the quality of the reamed hole in a cost-effective way. An important advantage of the designed simulation is the implementation of the first cut situation. Thus, not only the analysis of the steady state during the reaming process, but also the assessment of effects like the entrance of the cutting edges is possible. To make a comprehensive uncertainty simulation, a coupling of the chip cross-section simulation with a cutting force and tool stiffness model is necessary.

Simulation of the cutting forces for multiblade reamers

For the identification of uncertainty effects during machining processes the prediction and knowledge of the appearing cutting forces is an important aspect. For simulation, the force model is the connector between the chip cross-section and dynamic model during the reaming process, thus the key model. At present, different cutting force models, like Kienzle and Victor [6] for different machining processes or Chandrasekharan [10] for drilling operations, are used for this purpose. The modification of these models, e.g. Kienzle and Victor, with corresponding coefficients does not necessarily lead to desired results. Others, like Chandrasekharan, are very complex and have to be simplified for application to reaming processes. Furthermore, all known process force models neglect the important uncertainty of material strength gradients, which lead to large deviations in product quality of machining processes. Thus, a cutting force model for reaming with multiblade reamers is developed which includes this important uncertainty [11].

The resulting radial forces cancel each other out during an ideal reaming process with multiblade reamers. Thus, a model process for the acquisition of the load spectrum at a single cutting edge, consisting of cutting force F_c, passive force F_p and feed force F_f, is designed. Therefore, five of six cutting edges on the commercial reaming tools are removed. This makes the direct measurement of the load spectrum at a single cutting edge possible. The forces are measured by a three axis dynamometer, which is fixed on the machine table. Generally, the measurement setup can be build up in two different ways. The first one uses a standing tool which is directly fixed to the dynamometer and a work piece which is clamped by a tool holder to the machine spindle (kinematic reversal). This allows a direct measurement of the cutting forces, if the cutting edge is oriented in the coordinate system of the dynamometer. However, the test complexity and the time requirement is quite high because of the work piece sleeves which have to be changed after every test. The other test setup uses a standing work piece and a rotating tool (normal kinematic). Because of the extensive test matrix, this setup is used to reduce the time requirement. Due to the used construction, it is necessary to transform the measured forces F_x and F_y to the forces F_c and F_p. These are forces in the rotating coordinate system of the cutting edge. Therefore, a sensor is used which gives a trigger signal for every revolution. If the actual position angle α of the reference point and the relative distance angle δ between the reference and the cutting edge are known, the cutting force

$$F_c = \left(F_x \cdot \sin(\alpha + \delta) + F_y \cdot \cos(\alpha + \delta)\right) \cdot \left(\sin^2(\alpha + \delta) + \cos^2(\alpha + \delta)\right)^{-1} \tag{5}$$

and the passive force

$$F_p = \left(-F_x \cdot \cos(\alpha + \delta) + F_y \cdot \sin(\alpha + \delta)\right) \cdot \left(\sin^2(\alpha + \delta) + \cos^2(\alpha + \delta)\right)^{-1} \tag{6}$$

can be calculated with F_x and F_y.

To get a comprehensive cutting force model a large set of parameters is varied during the machining tests. The used tools are multiblade reamers with six blades and a diameter of 14H7. As mentioned above five of six cutting edges are recessed for the measurement of the process forces. This leads to a resulting radial force because no compensation of the resulting radial forces takes place during the cutting process. To minimize the uncertainty of the radial deviation during the tests, the unsupported length of the tools is reduced to achieve higher tool stiffness. Besides standard reamers, tools with geometric modifications (lead angle, flank width, backtaper ratio) are used.

These modifications are the same as for the machining tests with the skewness of the pre-drilled hole. The feed per tooth f_z and the cutting speed v_c are also varied. In addition to the former tests the width of cut is modified in three steps (0.05 mm, 0.1 mm, 0.15 mm). To involve the important uncertainty of the material strength gradient two different work piece materials with the same chip formation characteristics are used. The first one is heat-treated steel 42CrMo4V (1.7225) and the second one is case hardened steel 16MnCr5 (1.7131). The relation between the process forces and the varied process parameters are described mathematically by

$$F(a_p, v_c, f_z) = a_p^{m_{ap}} \cdot v_c^{m_{vc}} \cdot f_z^{m_{fz}} \cdot A, \qquad (7)$$

with the coefficients m_{ap}, m_{vc}, m_{fz} and A, determined by a multivariate regression using the empirical data.

The comparison of the cutting force F_c for 42CrMo4V and 16MnCr5 shows the influence of the material strength (Fig. 8). The cutting forces for 42CrMo4V are visibly higher than those for 16MnCr5. An enhancement in the feed per tooth f_z and the cutting depth a_p, causes an increase in cutting forces because of an increase in the chips cross-section. The variation of the cutting speed has different effects. On the one hand, the enhanced chip volume leads to an increase of the cutting force. On the other hand, the high cutting speed promotes the removal of the chips over the cutting face by a decrease in the friction coefficient. The negative coefficient m_{vc} shows the dominance of the reduction in the friction coefficient.

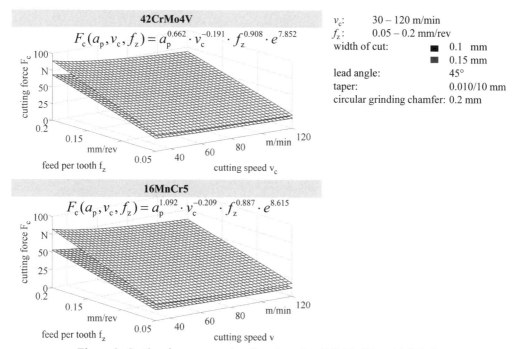

Figure 8: Cutting force regression functions for 42CrMo4V and 16MnCr5

To sum it up, the developed cutting force model for the reaming with multiblade reamers is an important aspect in the control of production uncertainties. The force model functions as the interface between the chip cross-section and static as well as dynamic models for process simulation. The consideration of different material strengths is an important improvement which, for the first time, allows the implementation of the uncertainty material strength gradient in future simulation models.

Conclusions and outlook

The presented methods for the multilevel process chain of drilling and reaming enable a systematic approach for the identification and control of uncertainty in production processes. The results of the tool benchmark show the possibility to reduce uncertainties by a smart tool choice. Further this test identifies the material strength gradient and the pre-drilled hole skewness, with a radial deviation up to 0.2 mm, as key uncertainties during the reaming process. So far, the influence of the skewness of a pre-drilled hole on the reaming process was not focus of research projects. Because of its relevance it was analyzed in further research. Contrary to popular belief, the results show that the reaming tool does not follow the pre-drilled hole for a diameter of 14H7 and a L/D-ratio of 8. Thus, the conventional guess that the reamer follows the pre-drilled hole has to be reconsidered. Furthermore, the results show that the simulation of the influence of uncertainties is an important aspect in uncertainty control. Therefore the simulation of the chip cross-section and resulting process forces are necessary. For this purpose, approaches have been presented too. A cutting force model was developed, which considers the uncertainty of a material strength gradient, for the first time. The cutting force model provides a means for the dynamic simulation of the reaming process to determine the uncertainty and the resulting product quality.

Based on the achieved results further research will be conducted. Thereby, a dynamic simulation model for the reaming process will be developed which considers the influence of the flank width. Furthermore, a voxel-simulation will be designed which allows a local variation of material strength. Thus, the analysis of the effects of a material strength gradient on the reaming process will be possible. The developed models also form a basis for an online identification of uncertainties using active machine components.

Acknowledgement

The authors would like to thank the German Research Foundation (DFG) for funding the research activities at the Collaborative Research Centre (CRC) 805 – Control of Uncertainty in Load-Carrying Structures in Mechanical Engineering.

References

[1]   F. Koppka: A contribution to the maximization of productivity and workpiece quality of the reaming process by analyzing its static and dynamic behavior, Shaker Verlag, Aachen (2009)

[2]   K. Ishikawa: Guide to Quality Control, Asian Productivity Organization, Tokyo (1992)

[3]   O. Bhattacharyya, S G. Kapoor, R. E. DeVor: Mechanistic model for the reaming process with emphasis on process faults, International Journal of Machine Tools & Manufacture, 46 (2006), page 836-846

[4]   O. Bhattacharryya, M.B. Jun, S. G. Kapoor, R. E. DeVor: The effects of process faults and misalignment on the cutting force system and hole quality in reaming, International Journal of Machine Tools & Manufacture, 46 (2006), page 1281-1290

[5]   DIN EN ISO 1101: Geometrical Product Specifications (GPS) – Geometrical tolerancing – Tolerances of form, orientation, location and run-out, Beuth Verlag, Berlin (2009)

[6]   E. Paukisch, S. Holsten, M. Linß, F. Tikal: Zerspantechnik (engl. Machining Technology), Vieweg+Teubner, Wiesbaden (2008)

[7]   F. Klocke: Manufacturing Processes 1 – Cutting, Springer-Verlag, Berlin Heidelberg (2011)

[8]   H.Schönherr: Spanende Fertigung (engl. Metal-Cutting Manufacturing), Oldenburg Verlag, München Wien (2002)

[9] E. Abele, T. Hauer. M. Haydn, C. Bölling: Reduced uncertainties during hole-finishing – New knowledge concerning the influence of pre-drilled holes on reaming process, wt Werkstattstechnik online, 101 (2011), page 81-87

[10] V. Chandrasekharan, S. G. Kapoor, R. E. DeVor: A Mechanistic Model to Predict the Cutting Force System for Arbitrary Drill Point Geometry, Journal of Manufacturing Science and Engineering, 120 (1998), page 563-570

[11] E. Abele, T. Hauer, M. Haydn: Modeling the cutting forces during reaming processes with multiblade reamers, wt Werkstattstechnik online, 101 (2011), page 407-412

Integration of Smart Materials by Incremental Forming

Matthias Brenneis[1, a], Markus Türk[2, b] and Peter Groche[3, c]

[1, 2, 3] Institute for Production Engineering and Forming Machines – PtU, Technische Universität Darmstadt, Petersenstraße 30, 64287 Darmstadt, Germany

[a]brenneis@ptu.tu-darmstadt.de, [b]tuerk@ptu.tu-darmstadt.de, [c]groche@ptu.tu-darmstadt.de

Keywords: Incremental Forming, Joining, Smart Structures, Rotary Swaging, Metal Spinning

Abstract. Today, the components of smart structures consisting of structural and smart materials are generally produced separately and assembled in additional processes afterwards. An alternative approach, which combines the forming of metallic parts and the assembly of the structures in one process step, is proposed in this paper. Incremental forming processes are applied for this operation. Significant joining mechanisms will be analyzed and some applications of this combined forming and assembly process are shown. As sensors, smart components allow a monitoring of appearing loads, as actuators they allow an active influencing on appearing disturbances. The research contains numerical analyses and experimental tests.

Introduction

This paper deals with the reduction of uncertainty of bar structures by integrating adaptive components into the bars. In this project, the foundations are laid for the integration of adaptronic components into hollow metallic structures by means of incremental forming methods. Results of the studies are technologies which allow for a safe coupling of actuators, sensors and bearing structures without causing the damaging of the actuators and sensors during the coupling. They feature the advantage that the forming of the parts as well as the integration of the active components can be realized in one process.

Joining by forming is widely used in industrial applications. Especially incremental bulk forming processes are suitable for joining operations. Rotary swaging for example is an established process for joining metal wires and bushings [1]. So far, the mechanisms enabling a high joining strength when incremental forming processes are applied, have to be identified. A deeper insight into the mechanisms would allow the determination of efficient forming strategies for the assembly of smart structures consisting of smart materials and metallic structures. The paper at hand is written to depict some of the phenomena.

Smart structures by incremental forming processes

Smart Structures. Smart structures are able to sense, control and actuate. They can be realized by composites consisting of smart materials (multifunctional) and structural materials (without additional functions) [2]. Compared to load carrying structures made of purely design materials, smart structures are able to react to uncertain external loads or structure behaviors with higher flexibility [3]. This is achieved by either appropriate structure modifications on the basis of monitored acting loads or a reduction of resulting stresses as a consequence of additionally induced stresses directed opposite to external loads.

Producibility of composites made of smart material and a metallic structure has been proven by joining processes like gluing or screwing and casting processes [4].

Incremental forming processes seem to be particularly attractive for the assembly of smart and structural components by enclosing the functional elements with a plastically deformable material. On one hand the assembly can be carried out without additional joining parts. On the other hand, a huge freedom of material selection is given since incremental forming processes can lead to a large formability. Further advantages result from the short clock cycles [5] and the possibility to vary the temperature increase of the workpiece by changing the process duration.

Incremental forming processes. Incremental forming is the oldest known technique in metal working. A general definition of this class of manufacturing processes is given in [4]: "In an incremental forming process, regions of the workpiece experience more than one loading and unloading cycle due to the action of one set of tools within one production stage." Differently to "die-defined" processes, where each material particle is deformed only once and the shape of the finished part is nearly totally defined by the geometry of the die, incremental processes create deformations within regions of the workpiece only. In incremental processes the product shape is determined by the kinematics of the tools. Incremental forming methods seem to be particularly attractive for this integration task because on one hand they manage the task without additional joining parts and are able to solve joining and forming tasks at the same time. On the other hand, in this case the forming and the local component load can be largely influenced by means of the selection of the tool paths.

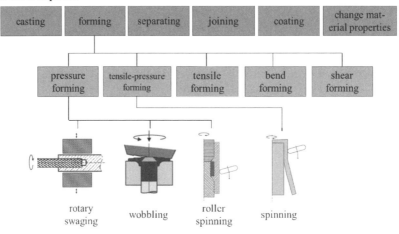

Fig. 1: Classification of incremental forming processes [6]

As to be seen in Fig. 1, incremental forming processes are classified in [6]. Due to the numerous advantages concerning material utilization and flexibility, several scientific investigations take place. For massive forming processes, this applies especially to rotary swaging [7], ring rolling [8], metal spinning [4], wobbling [9] and robot-assisted open die forging [10, 11].

In our researches two incremental processes are applied to join smart and structural materials: metal spinning and rotary swaging.

Integration of ring magnets by metal spinning

Metal spinning. Metal spinning is the transformation of a flat metal disc or a tube into a specific shape. It is achieved by rotating it at high speeds on a manual spinning lathe or performed by CNC controlled automated spinning machines. The tube is clamped at a spin block or 'mandrel' and a series of sweeping motions then evenly transforms the tube around the chuck into the desired shape. For the production of compound structures by metal spinning it is necessary to extend the established forming process by a relevant parameter [12]. Differently from the conventional process, where a single disc blank or a tube is formed, it is to be examined whether a ring can be formed into the tube during the spinning process in one step. Besides, it must be considered that the ring is made of a sensitive material. Therefore, the load on the ring has to be limited.

For the integration of ring magnets by metal spinning a design was developed, whereby two forming steps deform the aluminum tube into the desired geometry. Fig. 2 schematically shows the initial condition of the experimental setup on the top and the final condition on the bottom. In the first step the tube and the first ring magnet are positioned at the two-part mandrel and the tube is formed around the ring. After this, the second ring magnet is positioned on the mandrel and at the second step the final geometry is formed by the spinning roller. Now the mandrel can be removed and the two rings are fixed into the tube part.

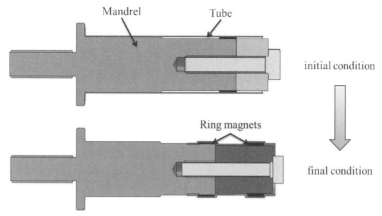

Fig. 2: Experimental setup for the integration of ring magnets by metal spinning

The produced cartridge with the integrated ring magnets can be used as sensors in bottom brackets of bicycles to monitor the cadence as well as the torque. Therefore, two Hall sensors are positioned at the axis below the ring magnets (Fig. 3, left). For the transmission, the sensor signal is boosted by operational amplifiers. Two sinus curves result from the rotation of the shaft. The interpretation of both signal curves gives information about current rotational speed and the angle of torsion of the shaft. The cadence can be identified by the frequency of the hall signal, the torsion of the axis by the phase displacement of the two hall signals. The produced bottom bracket with integrated sensors is shown on the right side of Fig. 3.

Fig. 3: Bottom bracket with integrated sensors

Integration of piezoceramics by rotary swaging

Rotary swaging. Rotary swaging is an incremental metal forming process for the reduction of cross-sections of bars, tubes and wires. Sets of two, three, four, or in special cases up to eight dies perform several simultaneous radial movements. Advantages of this process are: Different material combinations can be joined, the clock cycles are very short [13] and so they are economically reasonable [5].

During every inward movement of the dies, a small area of the workpiece is deformed. Depending on the kind of relative movement between workpiece and dies, rotary swaging is categorized into infeed and hitch-feed swaging methods [14]. Mainly compressive stresses inside the workpieces lead to plastic deformations during conventional swaging processes.

Integration of smart structures. To measure tensile loads via piezosensors, which can also be induced by bending moments, it is necessary to prestress the ceramic. Due to various applications, demands on process can differ in terms of strength of the piezoceramic, necessary pre-load and stiffness of the whole structure. Therefore, the influence of the ring geometry is focused.

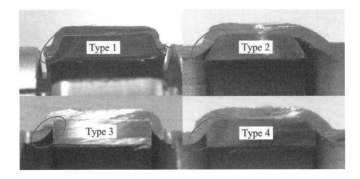

Fig. 4: Experimental tests of different ring geometries

Four different ring geometries have been determined. During the process, the ring is positioned at a mandrel. To evaluate the results, the tube with the integrated ring has to be separated after the process. At this, type 1 also shows the best result with a good contact zone between the ring and the tube. The ring is securely mounted into the tube. The other ring geometries show a gap between the tube and the ring. Fig. 4 shows the separated tubes to illustrate the contact zones of the different ring geometries.

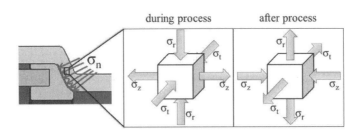

Fig. 5: Contact stress between tube and end cap (left); stresses during and after the forming process (right)

Contact stresses between ring and tube, which are responding for the axial prestress, are present due to spring back of the tube after its elongation in z-direction during the swaging process (Fig. 5). In absence of external forces, internal contact normal stresses can be seen as residual stresses caused by one or more of the four prestressing mechanisms:
1. Shrinking due to spring-back
2. Expansion due to spring-back
3. Thermal shrinking
4. Thermal expansion

The first two prestressing mechanisms are results of the elastic material response to load release. The importance of these mechanisms is well known for joining by forming processes. The resulting forces can be determined by using Hooke's law [15]. Thermal shrinking and expansion is inherent in every joining process induced by forming, because more than 90 % of the energy necessary for the forming process is transformed into heat. Therefore, a temperature increase in the plastically deformed zone occurs. Changes of contact normal stresses can be provoked by different temperatures inside the joining partners or by different coefficients of thermal expansion.

For even a proper quantification of the prestress state and how it is affected by the process, stresses and strains are investigated by means of a FE- analysis.

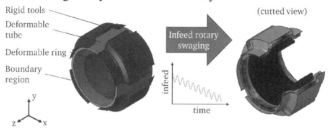

Fig. 6: FE- Model of tube and ring

Fig. 6 shows the modeling, where several modifications took place. To reduce the computing time in terms of numerical parameter studies, proper boundary conditions replace the tube adjacent to the joining site. Hereby, the count of finite elements is reduced. The outcome is a lower computing time. For further acceleration of the analysis a kinematic reversal takes place. Tools are moving in radial direction as well as rotating around the workpiece. Whenever deformable elements can be hold still, the numerical solver works more efficiently.

Result of the numerical investigation in terms of prestressing mechanisms is, if thermal shrinking is taken into account, the prestress is just 4% in excess of simulation without thermo-coupling. Hence, numerical simulations show that major prestressing mechanism of the process at room temperature is shrinking due to spring-back [16].

Besides the investigation concerning the prestress state, the non-destructive integration of the ceramic has to be regarded.

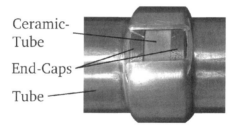

Fig. 7: Tube with integrated ceramic

Fig. 7 shows a sample consisting of an aluminum tube with a ceramic ring and end caps made from steel. The ceramic ring could be integrated non-destructively and is fixed by the prestressing in axial and the resultant friction-locking mechanism in tangential direction.

A crucial aspect for the production of smart structures by incremental forming is the susceptibility of the smart materials to damage. Basically, local stresses during assembly and forming can be significantly influenced by the selection of tool paths. Suitable paths can either be determined by careful process planning or by closed loop control of the local stresses on the basis of signals surveyed by the smart components during the manufacturing process.

Earlier studies showed that peak stresses in the smart materials can be reduced also by a suitable design of the interface between smart and structural materials [14]. Especially edges of the smart material have to be protected.

Conclusion

Smart structures offer an attractive opportunity to tackle uncertainties during usage with higher flexibility than passive structures. The necessary assembly of smart and structural materials can be achieved by incremental forming processes. Resulting joint strengths are mainly affected by the residual stresses after activation of positive locking. The crucial prestressing condition depends on shrinking and expansion of the involved components due to spring-back and thermal contraction. Finite element simulations with different physical models of the components allow the separation of the different mechanism and produced evidence of the activation of thermal and mechanical mechanisms. Examples of assembly by rotary swaging and spinning proved the existence of the investigated mechanisms. Furthermore, they demonstrated the producibility of smart structures by incremental forming without damaging of the sensitive functional components.

Acknowledgements

The investigations presented in this paper were carried out within the research project B4: "Integration of Functional Materials" of the Collaborative Research Centre 805 "Control of Uncertainties in Load-Carrying Structures in Mechanical Engineering" (CRC 805). The authors wish to thank the German Research Foundation (Deutsche Forschungsgemeinschaft DFG) for founding and supporting the CRC 805 and our research projects in the forthcoming years.

References

[1] S. Blum: Mechanical Rope and Cable, Report of the National Materials Advisory Board, Washington D. C. (1975).

[2] R. Platz, S. Ondoua, K. Habermehl, T. Bedarff, T. Hauer, S. Schmitt and H. Hanselka, in: Approach to validate the influences of uncertainties in manufacturing on using load-carrying structures, International Conference on Uncertainty in Structural Dynamics, Leuven (2010).

[3] V. Bräutigam, C. Körner and R. F Singer, in: Smart Materials by Integrating Piezoceramic Modules in Die Castings, Highlight Lecture, International Conference on Aluminium, Essen, Germany (2006).

[4] P. Groche, D. Fritsche, E.A. Tekkaya, J.M. Allwood, G. Hirt and R. Neugebauer:, Incremental Bulk Metal Forming, CIRP Annals - Manufacturing Technology 56 (2) (2007), p. 635-656

[5] P. M. Standring, in: Economic Aspects of Rotary Forging, International Conference – New Developments in Forging Technology, Stuttgart, Germany (2003).

[6] Deutsches Institut für Normung (engl: German Institute for Standardization): DIN 8580 Fertigungsverfahren (engl: Manufacturing Processes), September (2003)

[7] T. Rathmann and P. Groche, in: Development of a Technological Processor for 3D-FEA of Rotary Swaging Processes, 4th ICFG Workshop, Shanghai, China (2004).

[8] J. Allwood, et al.: The Technical and Commercial Potential of an Incremental Ring Rolling Process, Annals of CIRP Vol. 54/1 (2005), p. 233-236

[9] S. Rusz and M. Greger: New aspects of orbital forming technology, Engineering Plasticity from Macroscale to Nanoscale Pts 1 and 2 Vol. 233/2 (2003), p. 413-418

[10] T.J. Nye, A.M. Elbadan and G.M. Bone: Real-time process characterization of open die forging for adaptive control, Journal of Engineering Materials and Technology-Transactions of the ASME Vol. 123/4 (2001), p. 511-516

[11] O. Ziegelmayer, V. Schneider, R. Kopp and O. Durr: Analyses for the production of plane-curved work pieces by the assist of a robot in open-die forging, Materialwissenschaft und Werkstofftechnik Vol. 35/7 (2004), p. 447-453

[12] O. Music, O., J.M. Allwood and K. Kawai: A review of the mechanics of metal spinning, Journal of Materials Processing Technology Vol. 210 (2010), pp. 3–23

[13] Doege, E.; Behrens, B.A.: Handbuch Umformtechnik (eng: Forming Handbook), Berlin [u.a.]: Springerverlag (2007).

[14] M. Türk and P. Groche, in: Integration of adaptive components by incremental forming processes, SPIE Smart Structures/NDE, San Diego (2010).

[15] P. Groche and K. Tibari: Fundamentals of Angular Joining by Means of Hydroforming, CIRP Annals - Manufacturing Technology 55 (1) (2006), p. 259-262

[16] P. Groche and M. Türk: Smart Structures Assembly through Incremental Forming, CIRP Annals - Manufacturing Technology 60 (1), (2011), pp. 21-24

3. Uncertainty in Usage

FATIGUE LIFE ESTIMATION UNDER CYCLIC LOADING INCLUDING OUT-OF-PARALLELISM OF THE CHARACTERISTICS

Marta KUREK[1a], Tadeusz ŁAGODA[1b]

[1] Department of Mechanics and Machine Design, Faculty of Mechanical Engineering, Opole University of Technology,
ul. Mikołajczyka 5, 45-271 Opole, POLAND

[a] m.kurek@doktorant.po.opole.pl, [b] t.lagoda@po.opole.pl

Keywords: fatigue life, out-of-parallelism, fatigue characteristics

Abstract. The paper presents an algorithm of fatigue life determination for materials with no parallel fatigue characteristics under pure bending and pure torsion. The presented model uses the iteration method, and the applied fatigue criterion is function of the ratio of normal and shear stresses coming from bending and torsion, respectively. Three materials were applied for analysis: CuZn40Pb2 brass, 30CrNiMo8 medium-alloy steel and 35NCD16 high-alloy steel.

Introduction

A fatigue criterion under multiaxial loading is based on determination of an equivalent quantity which could allow to compare multiaxial loading with uniaxial loading. Many multiaxial fatigue criteria include a value of the following ratio in their equations:

$$B = \frac{\sigma_{af}}{\tau_{af}}, \qquad (1)$$

where: σ_{af} – fatigue limit for bending, τ_{af} – fatigue limit for torsion.

The criterion formulated by Gough and Pollard [1] belongs to that group of criteria. Gough and Pollard started wide research works on bending and torsion. Other criteria including a value of Eq. (1) were proposed by Nisihara and Kawamoto [2], Lee [3], Findley [4], Stulen and Cummings [5], Carpinteri and Spagnoli [6].

In the paper [7] Eq. (1) was presented for some constructional materials for which this relationship is constant. In such a case, fatigue life calculation is possible with use of the mentioned hypotheses. As for other materials, there is no one solution allowing the fatigue life assessment because variability of the following ratio must be taken into account:

$$B(N_f) = \frac{\sigma_a(N_f)}{\tau_a(N_f)}, \qquad (2)$$

depending on a number of cycles up to the fatigue failure.

In this paper, the algorithm of fatigue life assessment under proportional bending including out-of-parallelism of the characteristics was proposed with using Eq. (2). It is relevant in the case of materials which are characterized by out of parallelism of the characteristics such as CuZn40Pb2, 30CrNiMo8 and 35NCD.

The algorithm of fatigue life assessment

Stages of the algorithm of fatigue life determination under cyclic loading are shown in Fig.1. In the previous paper [8], a similar algorithm was proposed for the materials with out-of-parallel fatigue characteristics, but the results obtained for the loading combination were not satisfactory. The proposed model includes a change of the angle of the critical plane orientation which strongly influences the searched fatigue life. Like in the previous model, the fatigue life determination requires the input data, namely histories of components of the stress state tensors, according to the following equations:

$$\sigma_{xx}(t) = \sigma_a \sin(\omega t), \tag{3}$$

$$\tau_{xy}(t) = \tau_a \sin(\omega t - \varphi), \tag{4}$$

where: σ_a – amplitude of the normal stress coming from bending, τ_a – amplitude of the shear stress coming from torsion, ω – angular frequency, φ – phase shift angle, t – time.

In the presented algorithm, the normal stress history $\sigma_{xx}(t)$ relates to the stresses coming from bending, and $\tau_{xy}(t)$ concerns the stresses coming from torsion.
The next important step in fatigue life determination is determination of the critical plane orientation angle corresponding to the maximum effort of the material. In the paper, the critical plane position was determined with the method of damage accumulation, and the expression for normal and shear stresses was completed with the well known correction function using fatigue limits for bending and torsion [9, 10]:

$$\sigma_\eta(t) = \sigma_{xx}(t)\cos^2\alpha + \frac{\sigma_{af}}{2\tau_{af}}\tau_{xy}(t)\sin 2\alpha, \tag{5}$$

$$\tau_{\eta s}(t) = -\frac{1}{2}\sigma_{xx}(t)\sin 2\alpha + \frac{\sigma_{af}}{2\tau_{af}}\tau_{xy}(t)\cos 2\alpha. \tag{6}$$

In order to include variability of the parameter B depending on a number of cycles, the following relationships were derived:

$$\sigma_\eta(t) = \sigma_{xx}(t)\cos^2\alpha + \frac{B(N_f)}{2}\tau_{xy}(t)\sin 2\alpha, \tag{7}$$

$$\tau_{\eta s}(t) = -\frac{1}{2}\sigma_{xx}(t)\sin 2\alpha + \frac{B(N_f)}{2}\tau_{xy}(t)\cos 2\alpha. \tag{8}$$

In this paper, the criterion of maximum shear stresses [11] was used. It is strictly connected with the critical plane angle orientation, so the equivalent stress history can be written as:

$$\sigma_{eq}(t) = 2\tau_{\eta s}(t). \tag{9}$$

Relationships joining the fatigue life with the stress include the double-logarithmic relationship S-N according to ASTM [12] in the following form:

$$\log N_f = A_\sigma - m_\sigma \log \sigma_a, \qquad (10)$$

where: A_σ, m_σ – equation coefficients for bending, and

$$\log N_f = A_\tau - m_\tau \log \tau_a, \qquad (11)$$

where: A_τ, m_τ - equation coefficient for torsion

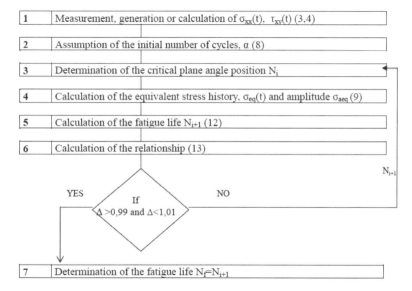

Fig. 1. Algorithm of fatigue life determination under cyclic loading

The fatigue life was determined according to the equation [12] after transformation of Eq. (10)

$$N_f = 10^{A_\sigma - m_\sigma \log \sigma_a}. \qquad (12)$$

The following ratio should be calculated:

$$\Delta = \frac{N_{i+1}}{N_i}. \qquad (13)$$

This procedure is repeated for successive calculated fatigue lives up to the moment when the following condition is satisfied

$$0.99 < \Delta < 1.01 \tag{14}$$

i.e. the error at the level of 1% was assumed – it is sufficient for fatigue calculations of machine elements and structures. If condition (14) is satisfied, the obtained fatigue life is the searched value.

Material and specimens

The following constructional materials were considered: brass CuZn40Pb2 [13], and high-alloy steels 30CrNiMo8 [14, 15] and 35NCD16 [16]. Smooth specimens of a round section were tested (Fig.2). Different values of the ratio $B(N_f)$ cause that it is not possible to apply the constant value of this ratio in the criteria including it. Table 1 contains coefficients of the regression equation for the selected materials according to ASTM standard [12]. Fig. 3 presents fatigue characteristics for one of the considered materials, namely 35NCD16. All the materials have out-of-parallel fatigue characteristics for bending and torsion.

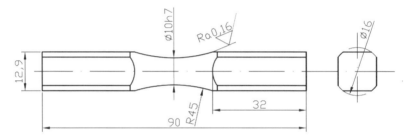

Fig. 2. Geometry of the tested specimens

Tab. 1. Coefficients of the regression equation for the selected materials

Material	Bending		Torsion	
	A_σ	m_σ	A_τ	m_τ
CuZn40Pb2	19.99	5.86	45.31	17.17
30CrNiMo8	27.54	8.05	69.58	24.62
35NCD16	31.95	10.03	44.51	15.08

Comparison of the calculated and experimental fatigue lives

The authors of the paper verified the test results for three constructional materials having non-parallel fatigue characteristics. Fig.4 shows comparison of calculated and experimental fatigue lives for brass CuZn40Pb2, and steels 35NCD16 and 30CrNiMo8. The graph includes also the situation when the parameter B=const (1). In such a case, the obtained results are not satisfactory and most of them are not included into the scatter band of the coefficient defined for bending.

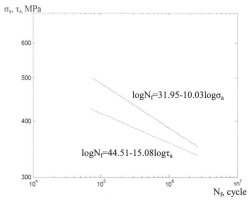

Fig. 3. Fatigue graph for pure bending and pure torsion for 35NCD16

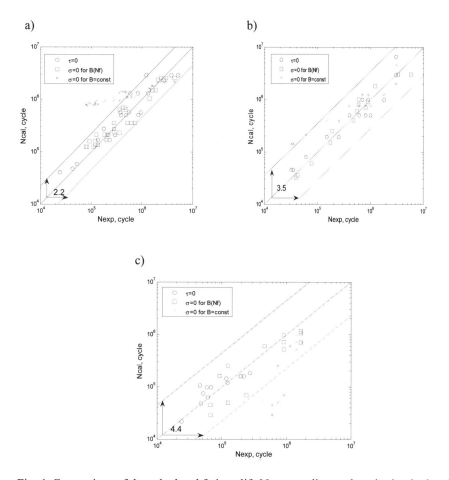

Fig. 4. Comparison of the calculated fatigue life N_{cal} according to the criterion in the plane of maximum shear stresses and the experimental fatigue life N_{exp} for pure bending and pure torsion a) CuZn40Pb2, b) 35NCD16, c) 30CrNiMo8

Comparison of the obtained calculated and experimental fatigue lives under combined bending with torsion for the selected materials is presented in Fig. 5.

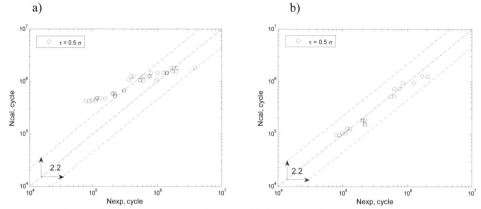

Fig. 5. Comparison of the calculated fatigue life N_{cal} according to the criterion in the plane of maximum shear stresses and the experimental fatigue life N_{exp} for $\tau_a = 0.5\sigma_a$ for CuZn40Pb2
a) B=const, b) $B(N_f)$

From the figures it appears that in the case of including variability of the coefficient $B(N_f)$ all the results are included into the scatter band of the coefficient 2.2. If B is constant, the results are worse.

Fig. 6. presents comparison of the calculated and experimental fatigue lives obtained under combined bending with torsion for 30CrNiMo8. It can be seen that the curve 6b shows better convergence of results when the ratio σ_a/τ_a (2) is dependent on a number of cycles.

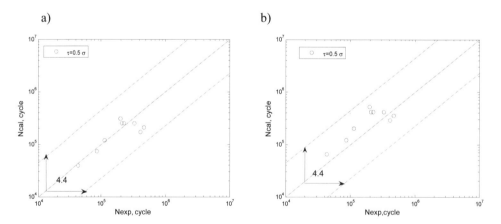

Fig.6. Comparison of the calculated fatigue life N_{cal} according to the criterion on the plane of maximum shear stresses and the experimental fatigue life N_{exp} for $\tau_a = 0.5\sigma_a$ for 30CrNiMo8
a) B=const, b) $B(N_f)$

The analysis shows that the presented algorithm is effective for CuZn40Pb2 than for steel 30CrNiMo8.

Conclusion

From comparison of the calculated and experimental fatigue lives it appears that the proposed algorithm for fatigue life assessment gives satisfactory results. The model can be applied for fatigue life calculations for materials with non-parallel fatigue characteristics under proportional loadings. An algorithm for fatigue life determination under non-proportional and random loadings is going to be formulated in future.

References

[1] Gough H.J.: Some Experiments on the Resistance of Metals to Fatique under Combined Stresses, London: His Majesty's Stationery Office (1951).

[2] Nishihara T, Kawamoto M.: The strength of Metals under Combined Alternating Bending and Torsion with Phase Difference. Memoirs of the College of Engineering, Kyoto Imperial University, Vol.X, No. 6 (1941)

[3] Lee S.B.: A criterion for fully reversed out–of–phase torsion and bending, Multiaxial fatigue ASTM STP 853, Philadelphia (1985), pp.553–568

[4] Findley W.N.: A theory for the effect of mean stress on fatigue of metals under combined torsion and axial load or bending, Journal of Engineering for Industry, (1959), pp.301–306

[5] McDiarmid D.L.: Fatigue under out–of–phase bending and torsion, Fatigue Fract. Engng Mater. Struct., Vol. 9, No. 6, (1987), pp.457–475

[6] Carpinteri A., Spagnoli A.: Multiaxial high–cycle fatigue criterion for hard metals, Int J Fatigue 23, (2001), pp.135–145

[7] Kurek M., Łagoda T.: Comparison of fatigue characteristics for some selected constructional materials under bending and torsion, 6th International Conference Mechatronic Systems and Materials, July, (2010) pp. 117-118.

[8] Kurek M., Łagoda T.: Algorytm oceny trwałości zmęczeniowej dla materiałów cechujących się nierównoległością charakterystyk zmęczeniowych w warunkach cyklicznego obciążenia (engl: Algorithm of fatigue life assessment for the materials with out – of parallel fatifue characteristics under cyclic loading), XXIV Konferencja Naukowa Problemy Rozwoju Maszyn Roboczych, Zakopane, (2011).

[9] Walat K., Łagoda T., Application of the covariance on the critical plane for determination of fatigue life under cyclic loading, Procedia Engineering, Vol. 2, 2010, pp. 1211–1218

[10] Walat K., Łagoda T., The equivalent stress on the critical plane determined by the maximum covariance of normal and shear stresses, Mat.-wiss.u.Werkstofftech., 2010, Vol.41, No. 4, pp.218-220

[11] Łagoda T., Ogonowski P., Criteria of multiaxial random fatigue based on stress, strain and energy parameters of damage in the critical plane, Mat.-wiss. u. Werkstofftech, (2005), Vol.36, No 9, pp.429-437.

[12] ASTM E 739-91 (1998): Standard practice for statistical analysis of linearized stress–life (S–N) and strain life (ε–N) fatigue data, in: Annual Book of ASTM Standards, Vol. 03.01, Philadelphia (1999), pp.614–620.

[13] Kohut M., Łagoda T.: Badania zmęczeniowe mosiądzu MO58 w warunkach proporcjonalnego zginania ze skręcaniem, (engl: Fatigue tests of MO58 brass under proportional bending with torsion), III Sympozjum Mechaniki Zniszczenia Materiałów i Konstrukcji, Augustów 1-4 czerwca (2004).

[14] Esderts A.: Betriebsfestigkeit bei mehrachsiger Biege – und Torsionsbeanspruchung, (engl: Fatigue under multiaxial bending and torsion), Fakultaet fuer Berbau, Huettenwesen und Maschinenwesen der Technischen Universitaet Clausthal, 9 Juni (1995).

[15] Sanetra C.: Untersuchungen zum Festigkeitsverhalten bei mehrachsiger Randombeanspruchung unter Biegung und Torsion, (engl: Studies on the strength behavior in case of multiaxial random loading under bending and torsion) Dissertation, Tech. Universitat Clausthal, (1991), p. 151

[16] Morel F.: Fatique Multiaxiale Sous Chargement D'amplitude Variable, (engl: Multiaxial fatigue at variable amplitudes) PhD thesis, Futuruscope (1996)

Approach for a consistent description of uncertainty in process chains of load carrying mechanical systems

Tobias Eifler[1,a], Georg C. Enss[2,b], Michael Haydn[3,c], Lucia Mosch[4,d], Roland Platz[5,e] and Holger Hanselka[2,5,f]

[1]Product Development and Machine Elements (pmd), Technische Universitaet Darmstadt, Magdalenenstraße 4, 64289 Darmstadt, Germany

[2]System Reliability and Machine Acoustics SzM, Technische Universitaet Darmstadt, Magdalenenstraße 4, 64289 Darmstadt, Germany

[3]Instuitute of Production Management, Technology and Machine Tools (PTW), Technische Universitaet Darmstadt, Petersenstraße 30, 64287 Darmstadt, Germany

[4]Department of Computer Integrated Design (DiK), Technische Universitaet Darmstadt, Petersenstraße 30, 64287 Darmstadt, Germany

[5]Fraunhofer Institute for Structural Durability and System Reliability LBF, Bartningstraße 47, 64289 Darmstadt, Germany

[a]eifler@pmd.tu-darmstadt.de, [b]enss@szm.tu-darmstadt.de, [c]haydn@ptw.tu-darmstadt.de, [d]l.mosch@dik.tu-darmstadt.de, [e]roland.platz@lbf.fraunhofer.de, [f]hanselka@szm.tu-darmstadt.de

Keywords: uncertainty, process model, process chains

Abstract. Uncertainty in load carrying systems e.g. may result from geometric and material deviations in production and assembly of its parts. In usage, this uncertainty may lead to not completely known loads and strength which may lead to severe failure of parts or the entire system. Therefore, an analysis of uncertainty is recommended. In this paper, uncertainty is assumed to occur in processes and an approach is presented to describe uncertainty consistently within processes and process chains. This description is then applied to an example which considers uncertainty in the production and assembly processes of a simple tripod system and its effect on the resulting load distribution in its legs. The consistent description allows the detection of uncertainties and, furthermore, to display uncertainty propagation in process chains for load carrying systems.

Introduction

Load carrying systems in mechanical engineering like suspensions, frames, etc. may be subject to different kinds of stresses, like mechanical, chemical, thermal stresses etc. They are designed to withstand especially mechanical stresses. If loads affecting the system are not sufficiently known, mechanical strength may be exceeded and the system may fail.

This study is part of the Collaborative Research Centre (CRC) 805: "Control of Uncertainty in Load Carrying Systems in Mechanical Engineering", which is publicly funded by the German Research Foundation (DFG). Within this scientific group, engineers and mathematicians formulated following working hypothesis describing uncertainty: *Uncertainty occurs when process properties of a system can not, or only partially be determined* [1]. The major aim of the CRC 805 is to first describe, then evaluate and finally control uncertainty. Uncertainty will be investigated along the process chain of product life, i.e. design process, production process and usage process.

This paper first presents the state of research for process descriptions and models to design, produce and use load carrying mechanical systems. Second, a new holistic approach to describe uncertainty along the process chain of product life is presented. Finally, an example of a simple load carrying tripod system is introduced to illustrate the benefit of the approach for the description of uncertainty.

State of research and motivation

The main purpose of load carrying systems is to sustain their load carrying capacity. Consequently, uncertain influences, e.g. inhomogeneous material, geometric deviations, higher loads than expected or external disturbances can affect the failure probability and may lead to severe safety-related and economic consequences. The occurring uncertainty thereby depends on a variety of influencing factors during the product life cycle as well as their interdependencies in and between processes. Therefore, a process model is essential for a consistent approach of uncertainty analysis.

In a variety of business, science and engineering applications, the visualisation of processes is a key step to clarify any kind of relations between input and output parameters. At the same time, few approaches offer detailed descriptions of technical processes. Cause-effect diagrams, such as the Ishikawa-diagram [2] and the process model of Heidemann from the field of product development [3] are based on categories of different influencing factors, e.g. from manual operations, machines, the environment etc. However, both approaches do not take into account propagation of uncertainty, e.g. manufacturing deviations may affect the load carrying capacity during usage. Furthermore, possibilities to reduce the level of uncertainty, e.g. by quality tests, are neglected.

To determine if other available modelling languages offer advantages for a consistent uncertainty analysis, approaches from the fields of business process management (BPM), software engineering, quality management (QM), as well as exemplary technological applications are reviewed in the following with focus on graphical visualisation. The aims are either an adjustment of existing business procedures or the identification of process requirements towards the supporting software. Especially the graphical representation of processes was formalised in different standards, e.g. structured analysis and design technique (SADT) or business process modelling notation (BPMN) [4]. Common to these approaches are different components of workflow diagrams to emphasise the relevant information flow, decision points and involved departments. However, other influencing parameters are usually neglected. The same applies to the QM process approach, which concentrates more on a system of decision points than on a single technical process, [5].

In contrast, approaches from the field of production management, such as value stream mapping [6], show the economical aspects in the shop floor. The process description thereby is limited to parameters such as cycle time, stock level, etc. and does not offer further advantages. Modelling approaches for technical systems in engineering also focus on specific problems, e.g. in control technology or thermodynamics [7, 8], and neglect the whole variety of disturbances that could affect load carrying systems.

Summing up, there is no comprehensive approach that accounts systematically for all influencing factors, such as ingoing product properties, disturbances, control steps, manual operations, machine accuracy, etc. Moreover, most technical process models disregard the numerous interrelations between subsequent processes. Usually, existing decision points or inadequate information are also not considered in combination with technical aspects. Especially for the analysis and control of uncertainty in load carrying systems, the cited approaches are not suitable. Consequently and based on literature, a new process model was developed within CRC 805 which will be presented in this paper. By means of the visualisation of technical processes and the description of influences, relevant factors for uncertainty can be identified and controlled. Thereby, the basic definitions of the terms process, state and state variable can easily be transferred. According to [5, 7, 8, 9], a process has an initial and final state and a finite duration. A process may carry out a conversion, transport or storage of information, energy and material, [5, 7, 9]. The properties of processes can be described by their time-dependent state variables, whereby the state is defined at a certain timestep [8].

Sample tripod system

Within CRC 805, uncertainty only occurs in processes, [1]. However, the consequence of uncertainty has effect on product properties of load carrying systems like scatter in geometry of parts, in material inhomogeneity or in resulting loads. To illustrate propagation of uncertainty within process chains, an example system concentrating on uncertainty in geometry is presented. Fig. 1a and 1b show a simple tripod system which distributes acting loads in three legs, assembled symmetrically with an angle of 120° by the connecting device. The idea of this tripod system is to demonstrate the effect of uncertain statistical deviations in production and assembly on real load distribution F_1, F_2 and F_3 in all three legs in usage processes compared to a desired equal load distribution $F_1 = F_2 = F_3$. The connecting device is a transparent cylinder with nine leg holes of three different diameters which can be assembled with the legs in three different ways. When loaded with an additional mass that leads to the force F, the force distribution in the legs in leg direction can be measured with force sensors and may differ due to tolerances in the assembly. A cap nut is mounted at the lower end of the force sensors, [10]. Fig. 1c and 1d show a simplified mechanical model of the tripod.

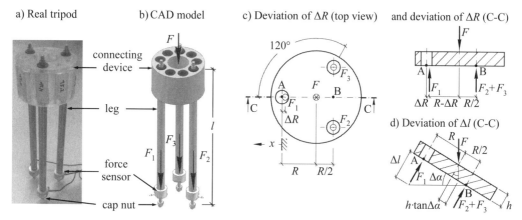

Figure 1: Tripod: a) real model, b) CAD model, c) schematic description of leg holes in connecting device, considering uncertainties of clearance ΔR in one hole and d) tilting of tripod, see [10]

In previous work [10], hole diameters in the manufacturing process of the connecting device are varied in an exaggerated way and allow to mount legs with a defined clearance in assembly, see Fig. 1c. The influence of this simulated, intentionally introduced uncertainty on the load distribution was analysed numerically. Experimental results in [10] show much higher scatter in load distribution according to variation of holes than simulated numerically. Therefore, in this paper, a further analysis of the system's production and assembly process is carried out and varying leg length is identified as an additional uncertainty. Thus, in this work, additionally the effect of varying leg length and consequently tilting of the tripod on the leg loads is taken into account numerically in a simplified way, see Fig. 1d. Key dimensions of the tripod are given later in this paper within the example.

An ideal symmetric tripod system with clearance $x = \Delta R = 0$ in all three holes, respectively, yields to leg loads $F_1 = F_2 = F_3 = F/3$, Fig. 1c. In [10], variation of hole diameter was examined. In this paper, only variation of one hole diameter with $\Delta R \neq 0$ is assumed, in which leg 1 is mounted. Thus, the leg may be mounted not centric but with clearance $\pm \Delta R$. According to Fig. 1c, posting equilibrium of moments in point B, leg load $F_1(\Delta R)$ yields to

$$F_1(\Delta R) = \frac{R}{3R \pm 2\Delta R} F. \qquad (1)$$

On the other hand, also length *l* of the first leg may vary. So tilting of the tripod correlates with the length variation Δl of leg 1. A schematic representation of the load conditions is shown in Fig. 1d. The used variables are variation of the leg length Δl, linearised small tilt angle $\Delta \alpha \approx 0$, mounting radius R of the legs in the connecting device, reference load F, leg loads F_1, F_2 and F_3 and height h of the force application point. Posting equilibrium of moments in point B, the resulting force $F_1(\Delta l)$ for the first leg depends on leg length variation Δl and yields to

$$F_1(\Delta l) = \frac{R - 2h \tan \Delta \alpha}{3R} F \cos \Delta \alpha \qquad (2)$$

with

$$\Delta \alpha = \arctan\left(\frac{2\Delta l}{3R}\right). \qquad (3)$$

Equations (2) and (3) enable to discuss uncertainty which occurs during three processes a) manufacturing and b) assembling and their effect on c) the load distribution during usage. The aim is to describe the propagation of uncertainties in this system in a consistent way in a process chain from a) to c). In the following chapter, a model of uncertain processes is presented in detail which allows consistent description of uncertainty within processes.

Model of uncertain processes

Generally, uncertainty in processes either leads to a product that does not meet the expectations after production or it results in unfavourable process output because of the product performance during usage. The example of the tripod system shows that component properties like geometric variations as well as a wide range of uncertain factors during assembly could affect the expected function. Varying hole diameters, realised on purpose, inhomogeneous material, geometric deviations of legs or the connecting device and assembly failures directly affect the load distribution in the three legs. External influences, like an uneven surface or vibrations, faults in manual operations, problems with assembly devices or the loading history may also have an effect.

For the assessment of decisive influences on the load distribution, a CRC 805 process model was developed and is used. By a consistent description of uncertain properties in the states and the influencing factors in the processes, it is a basis for a holistic approach of uncertainty analysis. Integral parts of the model thereby are states and processes. The process describes an action or an event that changes the system state, analogous to the definitions found in literature [5, 7, 9]. Properties of material, components or whole products are changed by the use of different appliances, e.g. forming, machining, assembly devices or the product during usage. The potential, external sources of uncertainty are grouped in four categories 1. disturbance, 2. information, 3. resources and 4. user, Fig. 2. A system boundary delimits the object of analysis, for example a single process, a chain of processes, a specific production step or the product usage. Moreover, the system boundary defines the analysed time span of a process, from time t_n in the initial process state to time t_{n+1} in the final process state one time unit n away. Additionally, the process model includes any kind of process monitoring or control, e.g. a function check of the tripod system during the usage processes, as well as a process number # indicating the level of detail of the analysis, e.g. the decomposition of processes into subprocesses and the sequence of processes, Fig. 5. With a more detailed analysis based on a decomposition of different process steps in production, more and most relevant uncertainties may be identified than without this approach.

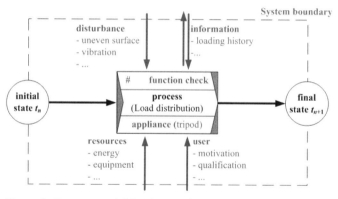

Figure 2: Process model for the consistent analysis of uncertainty

Matrix representation of state

The description of uncertainty in the different states of the process chain is a key issue for a holistic uncertainty analysis. It complements the consistent visualisation of processes. However, the description usually gets complex due to the large number of different properties. Thus, a systematical sorting of the properties into categories is needed. A review of common literature concerning design methodology [9, 11, 12, 13, 14, 15] shows that most properties of load carrying mechanical systems could be classified by six major aspects. These are a) geometric, b) material, c) mechanical, d) electrical, e) economical and f) additional properties, Fig. 3. For example, for the tripod system, aspects like length and diameter of the leg, diameter of the connecting device, bore diameter and assembly angle of the legs are grouped as geometric properties. Examples for the further categories are also listed in Fig. 3. The sixth category, additional properties, provides a higher flexibility concerning the addition of new aspects like safety values or extra information.

	geometry	kinematics	forces	dynamics	energy	substance	strength	material	substance prop.	chemical	electrical eng.	signal	information	technical data	safety	ergonomics	additional req.	heat engineering	economy	production	checkup	transport	transportation	usage	maintenance	recycling	costs	date	quantity
Pahl [11]	X	X	X		X	X						X				X	X		X	X	X	X	X	X	X	X	X		
Roth [12]	X	X		X			X		X	X	X					X				X	X								
Ponn [13]	X	X	X		X	X						X				X	X			X	X	X	X	X	X	X	X	X	
Naefe [14]	X	X	X		X	X								X		X	X			X	X	X	X	X	X	X	X	X	
Grote [9]	X	X	X		X	X						X				X	X			X	X	X	X	X	X	X	X	X	
Haberhauer [15]	X						X					X		X	X	X	X		X		X			X	X		X	X	X

derivation of property categories

a) geometric properties	b) material properties	c) mechanical properties	d) electrical properties	e) economical properties	f) additional properties
• length • diameter • angle • …	• material • chemical resistance • strength • …	• forces • moments • mass • mech. work • …	• el. tension • field strength • el. power • signals • …	• costs • quantity • duration • …	• safety • ergonomics • information • …

Figure 3: Criteria found in design methodology and derived property categories

Based on the concept of state vectors, found in literature [8], the product properties and the corresponding category of uncertainty can be summarised in a state matrix, Fig. 4. Whereas the relevant product properties are structured by means of the classification shown above, the corresponding uncertainty is described by three categories of the CRC 805 uncertainty model [16]. The level of uncertainty thereby can change with the available amount of trusted information from "Unknown Uncertainty", meaning that no statement is possible, to an enclosure of values with intervals in case of "Estimated Uncertainty" to a full description of "Stochastic Uncertainty" with known density functions. An essential aspect thereby is the indication of time t_n in the state matrix with n indicating a time step. A relatively long timespan of the process between initial state at time t_n and final state at time t_{n+1} usually leads to a less precise description of uncertainty, as the evaluation of all influencing factors and their effects on the process is at least difficult.

		Time	t_n		Uncertainty	
		Properties			Stochastic Uncertainty	Estimated Uncertainty	Unknown Uncertainty
	Attribute	Value	Unit			
Geometric properties	Bore diameter	$d+\Delta d$	mm	—	$[x_l; x_u]$	—
	Mounting radius	$R+\Delta R$	mm	$U(x_l; x_u)$	—	—
	Leg length	$l+\Delta l$	mm	—	—	?
	...	—	—	—	—	—
Material prop.		—	—	—	—	—
Mechanical prop.		—	—	—	—	—
Electrical prop.	Measurement accuracy	ΔF_1	N	$N(\mu; \sigma)$	—	—
	...	—	—	—	—	—
Economical prop.		—	—	—	—	—
Additional prop.		—	—	—	—	—

Figure 4: State matrix for the consistent description of uncertainty in states of process chains at a certain time t_n

Fig. 4 illustrates an example for one state of the tripod system in the beginning of the usage process at time t_n. On the left side of the matrix, properties are indicated by attribute, value, possible deviation and unit of measurement. The corresponding uncertainty on the right side of the matrix takes into account the whole time span of the process between time t_{n-1} before the current state and the current state at time t_n, e.g. geometric variation of all machined products or a whole load spectrum during specific usage processes. With regard to the analysis of the effect uncertain product properties might have on the load distribution during usage [10], the deviation of the diameter is described by a lower value x_l and an upper value x_u of the interval $[x_l; x_u]$, i.e. "Estimated Uncertainty". In comparison the uncertainty of the mounting radius of the legs in the connecting device, resulting from the specific assembly process, as well as the measurement accuracy of the used force sensors can be classified as "Stochastic Uncertainty". Whereas the deviation of the mounting radius follows a uniform distribution $U(x_l; x_u)$ between two values, a normal distribution $N(\mu; \sigma)$ with mean μ and standard deviation σ was adopted for the measurement accuracy. Further possibilities for the description of uncertainty are presented in [16, 17]. Although, it may influence the load distribution during usage, the variation of the leg length was not considered in the numerical analysis yet, leading to a state of "Unknown Uncertainty". Neither measurements nor numerical analyses were made for this aspect. For a consistent uncertainty analysis, the uncertain influences in the whole process chain, like positioning accuracy of machines, thickness of bonding in the system, temperature fluctuation of components, and their effects need to be evaluated more closely.

The consistency of the uncertainty description in the state matrix entails important advantages. The first is a consistent collection and representation of data. This simplifies the handling of large amounts of data and enables a reduced expense for the data processing. A second advantage is an easy comparability of the state matrices at different time steps among each other. Same state values are always added to the same category at the same line. For example this is important for the comparison of the load cases of the tripod system at different times or for varying usage processes. Also, reduction of uncertainty, e.g. by quality tests or fine machining, can be assessed easily.

Application of process model using the example tripod

As mentioned before, the assumed function of the tripod system is an equal distribution of leg loads $F_1 = F_2 = F_3$, Fig. 1. For an ideal case, the usage process leads to equal forces in all legs. However, measurements of the loads in the legs show scatter and therewith an unequal load distribution, [10]. Therefore, a detailed description of the process chain by means of the CRC 805 process model and a detailed characterisation of the product states are necessary to evaluate the uncertain influences in processes more closely.

Processes can be divided into hierarchical levels from high to low resolution. On the highest hierarchical level the process chain of the tripod system is divided into three main processes 1. manufacturing, 2. assembly and 3. load distribution, [10]. The rough classification into the basic processes of the products life cycle only allows a fast gathering of the process sequence and the associated states because of its low complexity. Since only the initial and the final state of component production, assembly and usage are taken into account, a detailed description of the interrelations between processes, uncertainty and effects on states is difficult or rather not useful. Thus, decomposition of the main processes is necessary for a more precise inspection of the global relation between the uncertainties in processes. This corresponds to a switch into the second or a lower hierarchical level. Fig. 5 shows the main process chain on the highest hierarchical level and a detailed decomposition into subprocesses of the assembly as well as the manufacturing chain on a lower level. The processes and states are illustrated without the belonging state matrix for better overview.

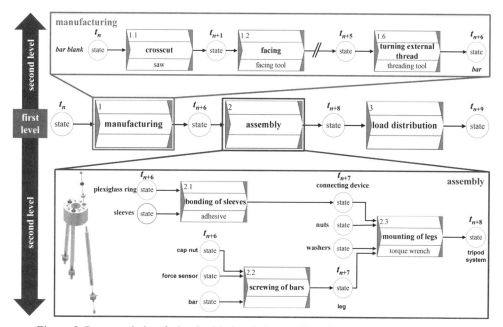

Figure 5: Process chain of tripod with detailed manufacturing and assembly processes

Thereby, the process number in the CRC 805 process model indicates the hierarchy level as well as the corresponding sequence of processes. On the highest hierarchical level, the three main processes for the tripod are distinguished, # = 1, 2 and 3 for manufacturing, assembly and load distribution. Every extension of the process number then gives information about the decomposition of these basic processes into lower levels. For example, process number # = 1.6 indicates the sixth subprocess of the manufacturing chain on the second level of detail. A further decomposition would be indicated, for example, by a process number # = 1.6.1 etc. For example, the manufacturing chain of the bar (part of the legs) consists of six processes within t_n to t_{n+6}. For reasons of clarity, only the first two processes crosscut (# = 1.1) and facing of face one (# = 1.2) as well as the last process turning of the external thread (# = 1.6) are shown in Fig. 5. In the analysis they are complemented by the processes longitudinal turning, cutting of the internal thread and facing of face two. All processes are connected by state matrices, see Fig. 4. The manufacturing example shows the possible representation of process chains with many subprocesses, like a sequence of manufacturing processes. Also the design of complex process chains with collateral processes, various initial states and merging of final states is possible, as illustrated in the assembly chain, # = 2.1 to 2.3 in Fig. 5.

For an extensive identification of uncertainties, a consistent procedure is necessary which allows a detailed screen of the whole process chain. This is a major advantage of the consistent description of uncertainties. For the tripod example, starting from the load distribution process, a review concerning the assembly and manufacturing is made, Fig. 6. Therefore, in the first step a search for all uncertainties, like the varying leg length Δl or the tilting angle $\Delta \alpha$ of the system, along the process chain is made. The second step is the analysis of the disturbances, which could lead to the identified uncertainties. Thus, an inspection of the subprocesses in the lower hierarchical levels is necessary. If the disturbances and the resulting uncertainties are known, the last step is the search for the linking of uncertainties along the process chain.

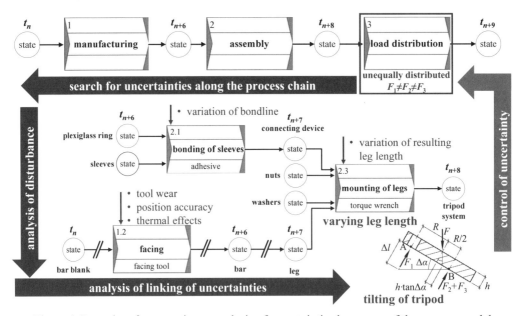

Figure 6: Procedure for a consistent analysis of uncertainties by means of the process model

For the tripod system, the detailed description of the usage process leads to the assumption that a varying leg length Δl has an influence on the load distribution. Thus, by a variation Δl of l the tripod system may be tilted by angle $\Delta \alpha$, Fig. 1d and Fig. 6. This leads to a variation of loads F_1, F_2 and F_3 in the three legs. Therefore, possible influencing factors are identified by means of the systematic description of the process chain, e.g. the bonding of sleeves during the assembly process. The usual thickness of the adhesive film is about 200 μm, but for a manual bonding, a variation of ± 100 μm seems to be possible. Further effects result from the manufacturing processes during the leg production. Influencing factors are the positioning accuracy of the machine tool, tool wear, etc. during machining processes, especially during turning of the external thread, Fig. 5, or thermal effects like the circadian temperature variation in the machine shop.

The uncertain influences in the processes of the manufacturing and assembly chain and their resulting effect on the product, i.e. the varying leg length Δl before usage, are shown in Fig. 7. Exemplarily, the state matrix summarises the uncertainty of product properties after the production of the tripod system. According to earlier analyses, the variation of the bore diameter $d = 15$ mm is described by an interval [15.0 mm; 15.4 mm], representing "Estimated Uncertainty". In comparison, a description by stochastic means, "Stochastic Uncertainty", is used for deviation of the mounting radius as well as for the measurement accuracy of the used forces sensors. The maximum value of the bore diameter leads to limits of mounting minimum radius $R_{min} = 34.8$ mm and maximum radius $R_{max} = 35.2$ mm. Due to the assembly processes, the variation of the mounting radius R follows a uniform distribution U(34.8 mm; 35.2 mm) around the nominal value $R = 35$ mm. For the deviation of the measurement accuracy, a normal distribution N(0 N; 0.4 N) with mean value $\mu = 0$ N and standard deviation $\sigma = 0.4$ N is given, [10]. For leg length $l = 235$ mm, the combination of production and manufacturing uncertainty along the process chain leads to an overall length variation $\Delta l = -0.119$ mm to 0.129 mm, see Fig. 7. This yields to tilting of the tripod system, i.e. a deviation angle $\Delta \alpha$ from the nominal value $\alpha = 0°$ between $\Delta \alpha = -0.13°$ to 0.14°, Eq. (3). As the variation of the influencing factors is only specified by a minimum and a maximum value, the uncertainty of deviating leg length l and tilt angle α is also represented by intervals, i.e. "Estimated Uncertainty", [234.881 mm; 235.129 mm] and [-0.13°; 0.14°]. The possible variation of the acting force $F = 44.15$ N and of its application height $h = 50$ mm are unknown. Fixed values are assumed for the evaluation of important influencing factors in the process chain.

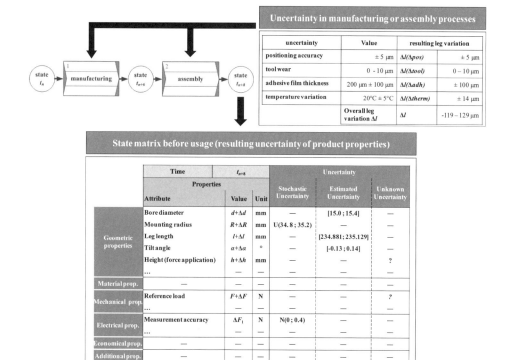

Figure 7: Uncertainty during manufacturing and assembly and effect on leg length before usage

Evaluation of effects

The deviations of mounting radius ΔR and leg length Δl are important aspects, their effects on the leg loads are analysed separately. On the one hand, a maximum deviation of the hole diameter $d = 15.4$ mm, already considered in [10], may lead to a varying mounting radius $R \pm \Delta R$, Fig. 7. With the maximum variation $\Delta R = 0.2$ mm, the resulting force of leg 1 is calculated with Eq. (1)

$$\Delta F_1 (\Delta R = 0.2 \text{ mm}) = \left(\frac{R}{3R - 2\Delta R} - \frac{R}{3R + 2\Delta R} \right) F = 0.112 \text{ N}. \tag{4}$$

On the other hand, a variation Δl of leg length l, Fig. 7, leads to tilting of the tripod system and thereby to an unequal load distribution. The resulting force in leg 1 is according to Eq. (2) and (3)

$$\Delta F_1 (\Delta l = 0.248 \text{ mm}) = \frac{R - 2h \tan \Delta \alpha}{3R} F \cos \Delta \alpha = 0.198 \text{ N} \tag{5}$$

with

$$\Delta \alpha = \arctan \left(\frac{2 \Delta l}{3R} \right) = 0.27°. \tag{6}$$

For the evaluation of influencing factors, the force variation $\Delta F_1(\Delta R)$ due to a mounting radius variation ΔR, Fig. 8a, is compared to the force variation $\Delta F_1(\Delta l)$ caused by leg length variation Δl, Fig. 8b. The comparison clarifies the significant influence of uncertainty by leg length deviation on the load distribution $\Delta F_1(\Delta l)$. From a mounting radius deviation of $\Delta R = 0.2$ mm results a force deviation $\Delta F_1(\Delta R) = 0.112$ N. The leg length variation $\Delta l = 0.248$ mm leads to $\Delta F_1(\Delta l) = 0.198$ N. As a consequence, the additional uncertainty due to leg length variation extensively covers the simulated uncertainty by mounting radius deviation and leads to a wide scatter of the measurement results.

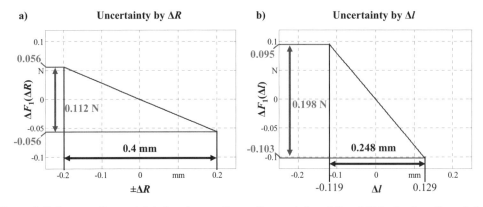

Figure 8: Influence of uncertainty by a) mounting radius variation ΔR and b) by leg length variation Δl on the load distribution

Furthermore, the systematic analysis of uncertainty along the process chain using the process model shows another important aspect. In special cases it is possible that uncertainties cancel out each other, so that measurements do not show uncertainty effects like load deviations. As a consequence, the conclusion could be drawn that no uncertainty exists. The description of the connection between the thermal deviation $\Delta therm$ and the positioning accuracy Δpos and their effect on the force deviation $\Delta F_1(\Delta l)$ clarifies this aspect, Fig. 9. The resulting leg length deviation Δl caused by $\Delta therm$ and Δpos mathematically can be described by

$$\Delta l = \Delta l(\Delta therm) + \Delta l(\Delta pos). \tag{7}$$

The combinations of $\Delta therm$ and Δpos, which lead to a length deviation $\Delta l = 0$ mm and as a consequence to a resulting force deviation $\Delta F_1(\Delta l) = 0$ N, are shown in Fig. 9. This effect underlines again the necessity of a process model for the systematic identification of uncertainty in process chains.

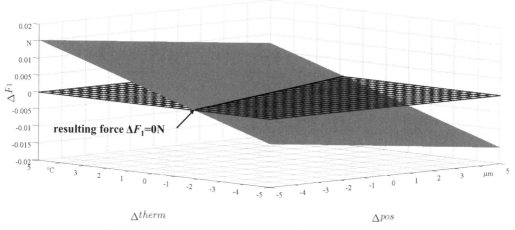

Figure 9: Possible cancellation of two different uncertainties

Conclusions and outlook

This paper presents a consistent approach for an evaluation of uncertainty in process chains. Based on a consistent visualisation of processes as well as state matrices to capture product properties, uncertain influences in process chains and their effect can be identified. The implementation in the analysis of a simple tripod system illustrates the representation of complex process chains and a consistent analysis of uncertainty and its propagation. For this example, the analysis of a tilt angle, caused by a deviation of leg length, shows a combination of uncertainties and a possibility to point them out by the usage of the process model. The deviation of leg length $\Delta l/l \cdot 100\% = 0.11\%$ thus results in a variation in the leg force $\Delta F_1/F_1 \cdot 100\% = 1.35\%$. In comparison, the deviation of the mounting radius $\Delta R/R \cdot 100\% = 1.14\%$ leads to a variation in the leg Force $\Delta F_1/F_1 \cdot 100\% = 0.76\%$. The relative influence of the uncertainty by leg length deviation on the forces is about twenty times higher than by the uncertainty of the mounting radius deviation. The results underline the importance of the consistent description and assessment of uncertain influencing factors within the process chain to evaluate such sensitive systems. Thereby, the level of detail of processes with regard to the hierarchy level as well as the analysed timesteps at t_n and t_{n+1} between two states can be adapted to the individual needs. Moreover, consistent description of processes and states facilitates electronic processing of uncertain data. Now, propagation of uncertainty in process chains can be described.

As an outlook, after a consistent description for process states has been presented, a description for time-dependent processes will be implemented. This will enable the process model to represent e.g. switching operations between different production processes with respect to economic uncertainties. Furthermore, control algorithms or wear mechanisms in usage processes may be implemented to consider technological uncertainties. For this, a problem specific description of the process is required. Eventually, a link will be established to an ontology-based information model for the exchange of uncertain data in load carrying structures. This will enable to have a communication interface between the single processes in a process chain.

Acknowledgements

The authors would like to thank the German Research Foundation (DFG) for founding this project within the Collaborative Research Centre 805.

References

[1] H. Hanselka and R. Platz: Ansätze und Maßnahmen zur Beherrschung von Unsicherheit in lasttragenden Systemen des Maschinenbaus – Controlling Uncertainties in Load Carrying Systems. Konstruktion (2010) No. 11/12, pp. 55-62.

[2] K. Ishikawa: Guide to Quality Control (Asian Productivity Organization, Japan 1992).

[3] B. Heidemann: Trennende Verknüpfung: Ein Prozessmodell als Quelle für Produktideen – Separating combination: A process model as a source of ideas for products. (VDI-Verlag, Germany 2001)

[4] R.K.L. Ko, S.S.G. Lee, E. Wah: Business process management (BPM) standards: a survey. *Business Process Management Journal* Vol. 15 (2009), pp. 744-791.

[5] DIN EN ISO 9000: Quality management systems: fundamentals and vocabulary, Beuth (2005).

[6] M. Rother and J. Shook: Learning to See: Value Stream Mapping to Add Value and Eliminate MUDA (Lean Enterprise Institute, USA 2003)

[7] H.D. Baehr: Thermodynamik - Thermodynamics, 12th ed., Springer, Berlin Heidelberg (2005).

[8] DIN 19226-1: Control technology; general terms and definitions, Beuth (1994).

[9] K.-H. Grote, J. Feldhusen (ed.): Dubbel – Taschenbuch für den Maschinenbau – Handbook of mechanical engineering, 22nd ed., Springer, Berlin Heidelberg (2007).

[10] R. Platz, S. Ondoua, K. Habermehl, T. Bedarff, S. Schmitt, H. Hanselka: Approach to validate the influences of uncertainties in manufacturing on using load-carrying structures. In: *Proceedings of USD2010 International Conference on Uncertainty in Structural Dynamics*, 20-22 Sep. 2010, Leuven (2010), pp. 5319-5334.

[11] G. Pahl and W. Beitz: Engineering Design – a systematic approach (Springer, Berlin Heidelberg, New York 2007).

[12] K. Roth: Konstruieren mit Konstruktionskatalogen Band 2 – Designing with design cataglogues, Springer, Berlin Heidelberg (2001).

[13] J. Ponn, U. Lindemann: Konzeptentwicklung und Gestaltung technischer Produkte – Concept development and design of technical products, Springer, Berlin Heidelberg (2007).

[14] P. Naefe: Einführung in das Methodische Konstruieren – Introduction to methodical design, Vieweg+Teubner, Wiesbaden (2009).

[15] H. Haberhauer, F. Bodenstein: Maschinenelemente: Gestaltung, Berechnung, Anwendung – Machine elements: Design, Computation, Application, Springer, Berlin Heidelberg (2007)

[16] R.A. Engelhardt, J.F. Koenen, G.C. Enss, A. Sichau, R. Platz, H. Kloberdanz, H. Birkhofer and H. Hanselka: A Model to Categorise Uncertainty in Load-Carrying Systems, in: *Proceedings of the 1st International Conference on Modelling and Management Engineering Processes (MMEP)*, 19. - 20. July (2010), Cambridge/UK.

[17] Sprenger, A., Mosch, L., Mecke, K., Anderl, R.: Representation of Uncertainty in Distributed Product Development, in *Proceedings of 18th European Concurrent Engineering Conference (ECEC)*, 18.-20. April (2011), London/UK.

Assessment of uncertainty for structural and mechatronics engineering applications

Stijn Donders[1,a], Laszlo Farkas[1], Michael Hack[1], Herman Van der Auweraer[1], Roberto d'Ippolito[2,b], David Moens[3,4,c], Wim Desmet[4,d]

[1] LMS International, Interleuvenlaan 68, B-3001 Leuven, Belgium

[2] Noesis Solutions, Gaston Geenslaan 11, B4, B-3001 Leuven, Belgium

[3] Lessius Mechelen University College, Department of Applied Engineering, J. De Nayerlaan 5, B-2860, Sint Katelijne Waver, Belgium

[4] Katholieke Universiteit Leuven, Department of Mechanical Engineering, Division PMA, Celestijnenlaan 300B, B-3001 Leuven, Belgium

[a] stijn.donders@lmsintl.com, [b] roberto.dippolito@noesissolutions.com, [c] david.moens@mech.kuleuven.be, [d] wim.desmet@mech.kuleuven.be

Keywords: uncertainty, vehicle subsystems, crashworthiness, fatigue life, virtual testing.

Abstract. Nowadays, mechanical industries operate in a highly competitive environment, therefore the process of developing a component from concept through detailed Computer-Aided Engineering (CAE) and performance validation is optimized for reduced development time and increased product performance. To continuously improve the product design and performance and reduce the costs and time to market, the design and performance engineering is shifted more and more towards virtual modeling and simulation processes from the expensive test-based design evaluations. Secondly, the booming introduction of active and adaptive systems in mechanical structures leads to a 'mechatronics systems' revolution, which further improves the product performance at the expense of increased system complexity. It is noted that the potential of structural dynamics test and analysis methods for addressing a structural dynamics design assessment or design optimization depends largely on the confidence that one can have in the results. That is, the results must be accurate, characteristic for the actual problem (and not be the result of testing artifacts) and representative for the actual behavior of the investigated structure. In this context, a key aspect is to be aware of the key sources of uncertainty in the designed product, and the impact thereof on the product performance in terms of structural dynamics, crashworthiness and/or acoustics. This paper reviews the main elements of test data and modal modeling uncertainty and assesses the impact of the uncertainty on some typical modeling problems taken from automotive and aerospace industry.

Introduction

In real-life structures, it is typically not possible to assign an exact value to all parameters [1]. Oberkampf [2] distinguishes two classes of non-determinism: *Variability* refers to the variation inherent to the physical system or the environment under consideration, while *uncertainty* is a potential deficiency in any phase or activity of the modeling process that is due to lack of knowledge. Other definitions may be agreed upon, but the above definitions are used in this paper.

In real-life structures, not all parameters are exactly defined. Non-determinism in input parameters results in scatter on the output properties that is not taken into account by a deterministic optimization. Therefore, fine tuning the deterministic input parameter settings without taking into account the real-life scatter on these parameters may result in a design for which the stability targets on the outputs are violated. When performing deterministic optimization, it should be taken into account that realistic uncertainty and variability on the inputs may propagate into scatter on the outputs. This, in turn, may lead to a violation of the constraints on performance targets. For this purpose, the robustness and reliability attained at the deterministic optimum should be assessed:

- Robustness is related to the sensitivity of the cost function to small changes in the inputs. A robust design is insensitive to the scatter in the input parameters.
- Reliability refers to the probability that a failure is attained as a result of input variability. A reliable design has a low failure probability with respect to pre-defined failure constraints.

Although the non-deterministic nature of the functional performance of mechanical/mechatronics structures is generally agreed upon, the authors note that non-deterministic modeling and simulation has not yet become a mainstream activity in vehicle development processes at OEMs and suppliers. The ever continuing advances in terms of IT resources are mainly being dedicated by analysts to increasing CAE model complexity, for which then only a limited number of deterministic simulations are performed; this holds for a range of CAE approaches, including Finite Element (FE) analysis, Computational Fluid Dynamics (CFD) and Multi-Body Simulation (MBS). This can be explained based on two key gaps in the state-of-the-art and state-of-the-use:
- Robust and reliability-based design optimization cycles typically require hundreds of deterministic analysis runs, so that the choice for this approach involves huge computational effort or a reduction of model fidelity (and consequently computational time).
- Quantification of (non-)deterministic data is a key challenge to include representative models of non-determinism in the design engineering process. Typically, insufficient (distribution) data is available, and the cost to collect enough data is too high (e.g. repetitive experimental studies on dedicated components and samples).

A traditional approach of taking into account the impact of design uncertainty is to include safety factors in the design, which however leads to sub-optimal designs with excessive weight and/or material usage. Better results can be achieved by introducing the non-determinism in the design process, and by taking into account its impact in the design process. Variability (material characteristics, manufacturing tolerances ...) can be handled with a probabilistic approach, while uncertainty (damping, boundary conditions ...) can be assessed with a possibilistic approach, e.g. fuzzy arithmetic. An overview of methods is provided in Section 1. Sources of uncertainties in different design engineering disciplines are discussed in Section 2. In Section 3, a number of application cases will be covered, taken from automotive and aerospace industry. More specifically, this comprises the study of a vehicle bumper subsystem subject to uncertainty, the robustness optimization of the fatigue life of a vehicle suspension subsystem and finally the controller design of a virtual shaker testing of a satellite with uncertain modal properties.

1. Non-Deterministic Analysis Methods in Computer-Aided Engineering

1.1 Reliability analysis and RBDO. After quantifying the variability of input parameters with statistical distributions, design criteria can be assigned and well-established Reliability Analysis [3] approaches can be applied to compute the system reliability. For practical purposes, reliability estimations are often expressed in multiples of the standard deviation Sigma: the larger the Sigma-value, the higher the system reliability. Nowadays, the Design for Six Sigma (DFSS) paradigm is often used: this aims at creating highly reliable 6σ designs (i.e. with a very small failure probability in the order of 10^{-10}). Two classes of methods are distinguished, which are worked out below:
- Reliability Analysis using Limit State Approximations;
- Reliability Analysis using Sampling Methods.

The first class of methods is *Reliability Analysis using Limit State Approximations*. In the design space, the boundary between safe and failing parameter combinations is often addressed to as the Limit State Function (LSF). The LSF can be approximated with linear or quadratic expansions at the nearest failure point (Most Probable Point, MPP), depending on the degree of nonlinearity. For a linear approximation, the First Order Reliability Method (FORM) [3] is generally used. The distance between the MPP and the nominal point is then a direct measure for the reliability index β. If a better approximation is needed, the Second Order Reliability Method (SORM) with a quadratic expansion can be calculated. Limit State Approximations (LSA) are very efficient and suitable when performance functions are approximately linear (which makes the approximation close to reality).

The second class of methods is *Reliability Analysis using Sampling Methods*, which is based on random sampling. The most straightforward is to perform a number of random simulations of the input parameters with the given distributions, and then to verify for each realization if this results in a failure. The actual failure probability can then be estimated by the ratio between the number of failures and the number of simulations. This is the standard Monte Carlo simulation approach [3], which is always applicable but has a very slow convergence rate (especially in case of small failure probabilities) when compared to the above-mentioned LSA methods.

However, Limit State Approximations are not always applicable. Particularly for highly nonlinear problems and/or problems with a high number of design variables, large errors may arise, and the computation times become prohibitive when using an LSA with a gradient-based sensitivity computation. For high-dimensional problems (number of inputs >15), alternative *Advanced Simulation Methods* have been worked out [4], with Line Sampling and Subset Simulation the most relevant. These dedicated sampling methods have a strongly improved convergence rate over standard Monte Carlo, and are able to solve high-dimensional reliability problems.

After assessing the reliability of a given design, the next step is to improve the reliability and/or performance and cost, by using a Reliability-Based Design Optimization (RBDO) approach [5]: constraints are defined on the required reliability, and the design is optimized such that all criteria (on cost, performance and reliability) are met. With an RBDO approach, an improved design can be found for the given input variability. An RBDO approach can be implemented as an optimization that targets to improve the reliability prediction of a FORM calculation; this strategy is denoted the *Reliability Index Approach (RIA)* for RBDO. Alternatively, one can use an inverse reliability formulation and applying the *Performance Measure Approach (PMA)* [5]. One then uses an inverse FORM approach with a target reliability level for each of the responses. The PMA formulation has been applied successfully to a number of structural cases and shows a very good performance and stability with respect to other available methods [5].

1.2 Fuzzy Finite Element (FE) analysis. Often in an engineering practice (especially in an early design stage), one is not able to quantify an input parameter in a probabilistic framework (type of distribution, mean, variance ...). One can then use a possibilistic description (between a minimum and a maximum, not larger/smaller than ..., etc) of the non-determinism in the inputs. Fuzzy arithmetic can be used to quantify these possibilistic descriptions and propagate the effect of the uncertainty on the system response, to obtain boundaries on this system response. This section first introduces fuzzy set theory [6] and outlines two strategies for fuzzy Finite Element analysis.

In classical set theory, the elements of a set either belong to the set entirely (membership level $\mu=1$), or do not belong to the set at all ($\mu=0$). This principle is generalized in *fuzzy set theory* [6]: a membership level $\mu_A(x) \in [0,1]$ is assigned to all elements x, so that the elements can belong to the set A to a certain degree. The *core* of the set is defined as the subset for which $\mu_A=1$. The *support* is the subset for which $\mu_A>0$ (also known as the *input vertex*). The *α-cut* (i.e. alpha-cut) is a generalized support: the subset for which $\mu_A \geq \alpha$. A *fuzzy number* is a fuzzy set with some specific properties: the set is convex and normal, the membership function is piecewise continuous and the core consists of a single element. An example fuzzy number is shown in Figure 1.

One can use fuzzy numbers to quantify the input uncertainty. The shape of the membership function can be derived from expert knowledge or practical measurements. Typically the *α-cut* strategy [7] is used to subdivide the fuzzy inputs into a number of intervals, as also shown in Figure 1. These intervals are then further processed to obtain interval (envelopes) predictions on the output responses. Two specific implementation strategies found in literature are enabling efficient fuzzy finite element analyses, resulting in applicability on industrially sized models [8]: the *Transformation Methods* and the *Optimization Approaches*.

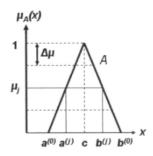

Figure 1: Fuzzy number: Example with a triangular membership function $\mu_A(x)$, with the core c and the support $[a, b]$. The α-cut strategy has been used to represent the fuzzy number as a set of 4 intervals $[a^{(j)}, b^{(j)}]$ at membership levels μ_j.

The *Transformation Methods* are a class of fuzzy methods that are based on Design of Experiments (DOE) methodology and conventional arithmetic.

Design Of Experiments (DOE) is a general approach to investigate complex correlations between input parameters and output response quantities [9]. The idea is to perform only a limited number of 'full calculations', which can be quite time-consuming in case of large industrial CAE models, and to use the results to obtain insight in the physics of the model and to perform subsequent design iteration and optimization. The selection of a DOE method depends on the available computational power and the expected order (linear, quadratic with/without cross terms ...) of the RSM model that is required to accurately represent the actual functional performances. A popular scheme is the full (or 2^n) factorial DOE scheme. For a problem with n input parameters, all 2^n possible combinations of parameter extrema (i.e. minimum and maximum values) are evaluated. Nowadays, DOE methodology is intensively used design space exploration, often combined with *Response Surface Methodology* (RSM) [10]: a meta-model of certain order is estimated from the experimental data, which yields insight in the functional relationship between the input parameters and the true response and output quantities. This small analytical model can then be used as a basis for further design iteration and optimization: the subsequent calculations are performed with quick analytical model evaluations, rather than full calculations.

The Vertex Method [7] is a fuzzy approach that performs a Full Factorial DOE on the input vertex of the fuzzy problem (the interval [a,b] at the lowest membership level in Figure 1). For n inputs, all 2^n parameter combinations are successively analyzed in a conventional deterministic calculation. The fuzzy output interval is then reconstructed from the deterministic results. The other members of the Transformation Method (TM) family are all extensions of the Vertex Method. The Reduced TM [11] first subdivides the inputs into intervals, and then applies a 2^n Full Factorial DOE for each of the intervals. For multiple inputs, the DOE plan will take the shape of a hypercube in the parameter space, in which all diagonals are evaluated. For comparison, the Vertex Method merely evaluates the corner points of the hypercube in the design space. Alternative methods have been presented, such as the General TM [11] (which adds more points and is thus better capable to deal with non-monotonicity at the cost of a high increase in CPU time) and the Short TM [12], which only samples a single diagonal in the design space. This is sufficient if the input uncertainty has a monotonic effect on the outputs (eigenfrequency values, eigenvectors, displacements ...). The TM often gives good insight in the actual uncertainty, but the method is not by definition conservative: one may miss response extrema if critical parameter combinations are not sampled and evaluated. Therefore, care should be paid to select a sufficiently large DOE size for the application of interest.

On the other hand, one can use *Optimization approaches* to perform a Fuzzy Finite Element analysis. The idea is to determine the lower and upper bound on the output of a classical finite element analysis by performing a search algorithm inside the domain defined by the interval inputs. If this search is successful (i.e., the global minimum and maximum of the analysis result are found), it returns the smallest hypercube around the actual solution set of the Fuzzy FE problem. As such, the solution of an interval finite element problem becomes an optimization problem, the goal function of which is the output of the deterministic FE analysis. The uncertain parameters are the design variables, constrained by the interval vector delimiting their ranges. The optimization is performed independently on each element of the result vector.

Contrary to the interval arithmetic approach, the optimization strategy approaches the interval result from the inside. This means that it does not guarantee conservatism, unless the actual bounds on the goal function are found. Furthermore, as the behavior of the goal function with respect to the uncertain parameters is rather unpredictable, the corresponding computational effort for finding the optima in general is strongly problem-dependent. Up to now, research on optimization approaches has focused on efficient procedures for performing the global search, using various approaches:

- **Global optimization** - The global optimization procedures perform the search for the exact bounds on the goal function by iteratively evaluating the goal function at designated points in the search domain. As one of the pioneers of fuzzy finite element modeling, Rao et al. apply a directional search based algorithm to tackle the optimization [13]. Other global optimization techniques often encountered in the framework of interval finite element analysis are linear programming and genetic algorithms [14]. Recently, the GαD-algorithm [15] has been introduced in the context of fuzzy finite element analysis.
- **Response surface methods** - In this approach, the goal function of the optimization problem is approximated by an appropriate surrogate *response surface model* (RSM) (Note that for 1-dimensional cases, with 1 input and 1 response, this takes the form of a response *function*). Subsequently, the optimization is performed on this RSM. The response surface methodology was first applied in the context of interval finite element analysis by Akpan et al. [16]. Key advantage is the avoidance of an exact goal function evaluation at each iteration step, which can be very costly. However, the accuracy of the approach relies completely on the exactness of the approximate response function.

For response-surface-based optimization, various strategies with different levels of complexity of the goal function approximation have been introduced. One can use the Vertex Method [7] and the Transformation Method [11] to sample the design space; for cases in which the goal function depends on the uncertain parameters in a monotonic way, this yields the exact solution. Recently, *adaptive response surface methodologies* have been introduced, which build the response surface during the analysis [19]. This reduces the number of function evaluations, by intensifying the sampling on the key part of the design space (i.e. near presumed locations of the response extrema).

The interval arithmetic procedure can be combined with the optimization approach in a hybrid interval procedure. In this context, an interval FE method to calculate envelope frequency response functions of uncertain structures was developed by Moens and Vandepitte [17]. The approach is based on the modal superposition principle, involving an optimization step in the modal analysis part of the procedure and an interval arithmetic approach for the actual superposition. Recent extensions have been made to enable the analysis of uncertain modal damping parameters, and to allow more efficient re-analysis [18] and optimization [19].

Possibility-based Design Optimization (PBDO) [5] is the fuzzy counterpart of RBDO. The idea is to perform an optimization on top of a fuzzy analysis: update selected design parameters to reduce the possibility of failure as outcome of the fuzzy analysis, thus resulting in a more conservative design in terms of cost and confidence level.

2. Sources of uncertainties in different design engineering disciplines

2.1 Automotive crash and safety simulations.
The vehicle crash event comprises a sequence of bifurcation-driven responses of parts that come into contacts, involving ruptures (failures of parts, connections), and with each sub-system response highly dynamic and non-linear [20]. Vehicle crash is a highly transient dynamic event, and even small changes on the input parameters can have a substantial impact on the crash behavior. Accordingly, it is of primary importance to know the most

sensitive parameters in the crash analysis and to assess the impact of uncertainty in these parameters on the performance of the structure in the crash event. For this purpose, key uncertainties must be accounted for already in the virtual design phase, so that the assessment can take place in a virtual environment, and virtual optimization can be used to improve the design. Several modeling challenges still exist in the automotive design for crashworthiness, concerning material behavior, modeling of connections (e.g. spot-welds). Such difficulties arise in finding accurate failure models, accurate capturing of strain-rate and temperature-dependent material behavior, and modeling of plastics and foam materials, which leads to model non-determinism.

In safety simulations implying virtual crash-test-dummies, several factors need special attention for a realistic dummy response: dummy positioning, correct restraint-systems and sub-assemblies interaction capturing, foam and other key material modeling [21]. Due to uncertainty in these factors, and due to the inherent and uncontrollable numerical noise, the safety simulations obtain a non-deterministic nature. As a natural consequence, the virtual certification tests imply the usage of compliance corridors and acceptance windows. Moreover, the validation process becomes more demanding and time-consuming in both the crash and the safety simulations.

Another non-deterministic factor that influences the robustness of the results is introduced by the numerical scatter that is inherent to computer implementation of numerical algorithms. Crash prediction algorithms based on the explicit time integration scheme have to deal with approximation errors [22]. The repetitive time-step integration sums up these errors (and the round-off errors) into the numerical noise which has to be small in comparison to the scatter due to model-parametric uncertainties. Non-linear contact problems with multiple bifurcations in combination with numerical noise result in large response variation without significant physical parameter variation. Numerical scatter is influenced by the choice of contact models, failure models, strain-rate effect in non-linear material model, element formulation, damping, mesh quality. This aspect, together with the natural existence of semi-controllable (at increased cost) and controllable model-parametric uncertainties such as dimensional tolerances, material parameters, positioning and initial conditions emphasizes the need of uncertainty assessment in automotive crash and safety simulations. Section 3.1 presents a study of a vehicle bumper subsystem subject to model uncertainties and their effect on the crash performance. Furthermore, the added value of the uncertainty-based design optimization will be illustrated in a bumper subsystem optimization study.

2.2 Mechatronics system design. One of the major evolutions that is currently taking place in mechanical industry is the increase of the electronic and mechatronic content in order to push the performance upward, and hence the product value and reliability. This is the case for the automotive as well as aerospace/aeronautics sectors [23]. As a result, the industry is facing no longer the mere challenge to design and develop a mechanical structure, but has to deliver mechatronics structures, and adopt the design, development and validation procedures to the changing needs of their final products. In terms of reliability engineering, the mechatronics evolution has a dual impact:

- On the one hand, there's a drastic increase in the system complexity, with additional sources of model uncertainty and subsystem failure possibilities. This calls for a systematic analysis of the mechatronics system model, a sensitivity analysis on possible sources of uncertainty, and the development of systematic failure mode and effect analysis methods to understand the chain of events from subsystem failure up to system-level performance failure.
- On the other hand, the introduction of control systems allows new strategies to mitigate the effect of model uncertainty and subsystem failure by means of control strategies to compensate for a given uncertainty or to switch to an alternative mode of operation that maintains the system performance at an acceptable level.

The booming introduction of active and adaptive subsystems is irreversible, in that it allows industry to push higher the product performance bar by optimizing not merely the mechanical performance but instead the mechatronics system performance of the products. For instance in vehicle industry, active systems (ABS, ESP, active suspension, active steering...) are key to increasing the vehicle comfort and safety of new products that are brought to the market [23]. However, automotive manufacturers must ensure the reliability of the mechatronics systems, since each failure that occurs will be put under a magnifying glass in terms of media attention. This has been seen in recent years in the media (see e.g. [24]), with reports on mechanical failures in new electronics components, causing vehicles to turn inactive now and then. This potential negative 'branding' in case of failures pushes the bar very high for industry in its quest to guarantee the system reliability of its mechatronics products.

Uncertainties in mechatronics systems can be distinguished on several levels:
- Sensor and Actuator level;
- Data acquisition and post-processing level;
- Cascading from subsystem to system level.

Uncertainties that are present on the level of sensors and actuators introduce errors that form the subject of many international standards (see e,g, DIN-EN-ISO 12100–1 or ISO 5725). These uncertainties need to be accommodated by the control, or for which robustness must be built into the system design [26]. These can be deviations from the nominal linear parameters such as sensitivities and gain or phase factors in all sorts of components and subsystems (material parameters, electric gains, filter cut-off frequencies…), but also very important is that most of these mechatronic interface components are to some degree nonlinear. This is in particular the case for the actuator systems [29]. Even when this nonlinearity is seemingly included in the applied models, this is typically a functional approximation that leaves part of the actual nonlinear behavior unmodeled. The propagation to the actual system level behavior may even be that the controller, designed based on the linear or simplified system model, becomes unstable. Hence the validation of the robustness of the control functional behavior under deviations of the actual system model with respect to the design model is very important.

Further complexities in the mechatronics systems approach may be related to the fact that inputs needed by the control system are not readily available from direct measurement. This leads to the need for a state-estimation procedure (and subsystem in the controller), which will be typically designed for uncertain system parameters as well as for measurement noise [26]. For example, measurement noise on a displacement signal which needs to be differentiated to a velocity signal may be significantly subject to amplification. Non-modeled nonlinear system behavior may distort the state estimator as well. Kalman filters may be adopted to balance the effect of measurement and the modeling errors. But also intrinsic uncertainties may influence the state estimation process, for example the unknown weight and inertia distribution in a car due to driver, passenger and luggage. Insufficient measurement data is typically available to correctly identify these properties and to use them in a state estimation procedure, hence the use of optimization processes and the need for robust control methods. Naturally, also mechanical uncertainty remains an important aspect for understanding the uncertainty of mechatronics systems. This comprises uncertainty in subsystem material properties and connectivity, modal mass and damping properties on a (sub)system level.

2.3 Modal analysis. Experimental Modal Analysis is nowadays a widely accepted methodology for the analysis and optimization of the dynamic behavior of mechanical and civil structures. Modal analysis is a technique that investigates and estimates parameters that describe the dynamic behavior of a mechanical structure under dynamic excitation, in terms of resonant frequency, mode shape and damping. Nowadays, Modal tests are a standard part of the analysis and refinement of physical prototypes or even operational structures. In Experimental Modal Analysis (EMA), a known excitation is applied to the structure, whereas Operational Modal Analysis (OMA) refers to investigating the dynamic behavior of a structure subject to an unknown excitation (e.g. wind loading on a bridge). For additional details on modal analysis, an interested reader is referred to [27] and [28]. After a modal analysis has been conducted, the obtained deterministic modal model can be used for verification of the fulfillment of the design criteria, for validation and updating of CAE models and for integration in hybrid system models. In reality, the modal results are just an

estimation of the model parameters based on a series of input-output or output-only tests and hence subject to testing and modeling errors, either stochastic disturbances on the input/output data, or caused by invalid model assumptions or data processing effects. Some key sources of errors are:
- Sensor location and orientation errors;
- Test set-up loading and constraining effects;
- Sensor loading effects on the test structure;
- Sensor calibration and data conversion errors;
- Disturbance and distortion in the test data measurement chain;
- Signal processing errors;
- Model estimation errors.

These error sources as well as the commonly used approaches to reduce these are briefly reviewed in [25]. Further application aspects are covered in [27]. As all modal tests are to some degree subject to such errors, the modal parameters can only be known within an uncertain interval.

2.4 Fatigue analysis. Classically for fatigue, the influencing factors are categorized into
- Loads;
- Material;
- Geometry.

But there are many more external factors than mechanical loads, like temperature influences and environmental influences. The geometry denotes the final design of the component including the tolerances in the manufacturing process. All other internal factors like influences of the used manufacturing process, surface treatment, changes in the quality of the material charge, etc are categorized under the material. A measure for the scatter in fatigue resistance of components is described in the ratio of load at 90% probability of failure to the load at 10%. Typical influences on the component side [37], like scatter in welds give a ratio of about 1.4 in the fatigue prediction, whereas scatter in base material accounts for about 1.15, and manufacturing tolerances for about 1.05. Compared to this, a typical scatter of the loads leads to a ratio of 2. The example in Section 3.2 shows the influence of the load transfer by the different components.

3. Application Examples

3.1 Fuzzy optimization study of a vehicle bumper subsystem subject to uncertainties. Bumper systems play an important role in the energy management of vehicles during low-speed accidents. Optimization technology and automation enables efficient balancing between different performance attributes. Optimization is typically done without considering the different model uncertainties such as properties subject to tolerances, environmental effects, non-uniform material properties, etc. The possibility-based design optimization (PBDO) process that implies fuzzy finite element (FE) analysis allows the designer to include the effect of non-deterministic model properties in the design performance optimization of the bumper subsystem [30]. The bumper subsystem optimization based on this rationale results in product performance that is guaranteed in the presence of uncertainties already from the early design stages.

An industrially representative FE bumper assembly has been selected (see Figure 2) to demonstrate the use and the added value of the PBDO methodology in the design optimization process. The bumper is subject to two different crash load cases: the Allianz reparability test AZT [31] and the pole-impact test; the latter approach has not been standardized, but it is extensively used by automotive OEMs in order to identify stiffness issues in their bumper design and to assess vehicle crashworthiness to a pole impact [32]. For the bumper assembly, five different uncertain design parameters have been defined: 3 geometrical dimensions and 2 shell thicknesses, representing production tolerances. The effect of the range of the uncertain bumper parameters is evaluated for the Allianz crash load case. The sectional force across the vehicle longitudinal beams (represented by rectangular extruded shapes) is the total force acting across the beams in the direction of the impact. The uncertainty effect on the longitudinal beam sectional normal force is shown in Figure 3. Here, the force has been measured in the x-direction (i.e. the direction $-v$, with v the impact velocity of the bumper system onto the barrier, i.e. parallel with the vehicle longitudinal beams. The curve indicates the energy absorption capability of the bumper: the area under the curve is the equivalent to the initial kinetic energy that is transformed into deformation energy. Important design constraint for this particular load case limits the largest normal sectional force which is correlated to the deformation length.

Figure 2: FE mesh and assembly of the bumper system.

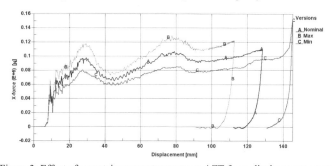

Figure 3: Effect of uncertain parameter range on AZT force-displacement curve.

The largest deformation length is a characteristic of the geometry of the bumper design equal to approximately 140 mm. The red curve ("C") indicates that the design based on the lowest parameter values has already reached the full energy absorption potential. This is clearly indicated by the peak in the normal sectional force reaching over 14 kN value after the full deformation potential of the bumper section has been consumed. It can be concluded that typical manufacturing tolerances have a large influence on the design response, which must be considered for a robust and optimal design.

The bumper is optimized for weight (quantified by the variable "Mass") and AZT force-displacement curve profile (quantified by the variable "RMSE F_x,") while applying a constraint for each load case. The central intrusion (quantified by the variable "Max Int") is limited for the pole-impact load case and the largest cross sectional normal force (quantified by the variable "Max Fx") is limited for the AZT load case. These two variables are acting as constraints in the optimization process. Two optimization strategies are compared:

- Classical *crisp* (i.e. *non-fuzzy*) *optimization* with uncertainty assessment of the optimum;
- A *PBDO-based optimization* process that takes the effect of parametric uncertainties into account, such that the system level of failure possibility is acceptable. The possibility of failure is function of the degree of constraint violation.

Figure 4 compares the results of two optimization scenarios. The first scenario based on a classical crisp optimization is compared to the second optimization scenario considering model uncertainties. The uncertainties modeled with triangular membership functions based on the concept of fuzzy numbers are propagated by the means of the fuzzy FE method, using the Reduced Transformation Method (See Section 1.2). It is important to note that the crisp optimum converges to a design point that violates the constraint applied on the AZT load case. This is indicated by a red vertical line in Figure 4 a), in the figure that shows the fuzzy result of the AZT constraint "Max F_x". This demonstrates that the PBDO-based optimization approach leads to an optimized design performance that is guaranteed in the presence of uncertainties.

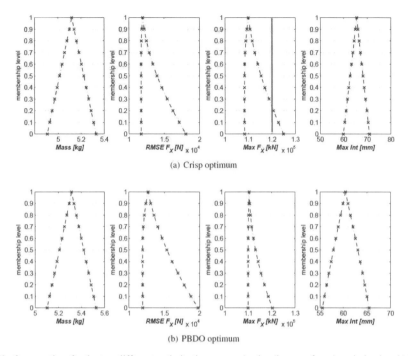

Figure 4: The fuzzy optima for the two different optimization cases: a) crisp (i.e. non-fuzzy) optimization, b) PBDO.

3.2 Robustness optimization of the fatigue life of a vehicle suspension system

To explain the methodology used to assess the design robustness and reliability, a typical vehicle dynamics application case has been considered. In particular, a model of a suspension system has been taken into account (see Figure 5). The model considered for the analysis includes three components: a knuckle, a tie rod and a lower arm. The suspension subsystem is part of a complete vehicle that has been analyzed using a multi-body simulation. This simulation computes the loads and the reaction forces generated by simulating the full vehicle run along a virtual track, representative of the real terrain profile and typical of the operational condition of the vehicle. The loads and the reaction forces computed in this way have been then used as input load histories for the assessment of the accumulated fatigue damage in the three subcomponents. The computation of the fatigue damage accumulated has been carried out by using a combined FE-MBS approach.

Figure 5: Multi-body model of the suspension system with tire.

Design Variables. In order to characterize the variability of the durability performance, 6 design parameters have been identified and characterized with a probabilistic approach. This statistical characterization is representative of the real variability due to geometrical tolerances, material properties and vehicle configurations. That is, 3 geometrical, 2 material properties and 1 vehicle mass configuration parameters have been considered, as listed in Table 1.

Table 1: Input design parameters: all Gaussian random numbers (normal distribution)

Name	Distribution	Mean	Standard Deviation
Knuckle Fillet Radius	Normal	Undisclosed	0.4 mm
Lower Arm Thickness	Normal	Undisclosed	0.4 mm
Tie Rod Radius	Normal	Undisclosed	0.4 mm
Tie Rod Tensile Strength	Normal	Undisclosed	20 MPa
Knuckle Tensile Strength	Normal	Undisclosed	20 MPa
Vehicle Mass Scale	Normal	Undisclosed	0.0025

These design parameters have been used to perform design changes, based on their statistical characterization. An automatic geometry meshing process has been created to take into account design changes in the geometry of the sub-components. Moreover, in order to limit the total computational effort required for a full optimization process, a hybrid optimization approach has been used, taking advantage of the Design of Experiments (DOE) and the Response Surface Modeling (RSM) techniques. The approach used can thus be summarized in the following steps:

- Design space exploration with DOE. As DOE, a Latin Hypercube (LH) [9] sampling of 40 samples has been performed to evaluate the design space spanned by the 6 variables.
- Response surface modeling of the functional performance. The type of RSM model has been decided based on an RSM study of several models; the best accuracy has been obtained with a Radial Basis Function (RBF) with linear Splines [35].
- Reliability assessment and optimization, based on the response surface model
- Validation of the obtained results.

For each component, the maximum absolute stress value has been computed. The maximum of these three values has been taken as the system performance for the present case. The robustness of this maximum overall value was assessed and optimized if it doesn't satisfy the requirements.

Results of the Reliabilty Analysis & RBDO. First, a reliability analysis using the Reliability Index Approach (RIA) has been carried out. Using the results of the DOE plan, a Response Surface Model (RSM) has been computed to approximate the FE model response in terms of the maximum absolute stress over the three subcomponents of the suspension system. The RSM has been used as a meta-model in order to assess the reliability of the suspension. This first reliability analysis step was necessary to find the position of the Most Probable Point (MPP). This was located at a reliability index of -0.36σ, corresponding to a probability of failure of 0.64 (i.e. 64%). The reliability of the design is measured in the number of 'sigma steps' between the nominal design and the MPP; for a reliability-based design, typically a reliability index of at least 3σ is required; often a higher value up to 6σ is imposed [3]; this will correspond to a sufficiently low probability of failure. To assess the effect of the introduced variability on the system response, a probabilistic characterization of the maximum absolute stress for the original design point has been carried out. In order to estimate the variability in the maximum absolute stress, an efficient Monte Carlo simulation has been performed, in which the RSM has been used as an efficient surrogate model for the full analysis chain. As estimate for the maximum absolute stress, this yielded a Gaussian random variable with mean value of 611 MPa and a standard deviation of 25 MPa.

Subsequently, a Reliability-Based Design Optimization (RBDO) has been performed. The target of the RBDO analysis has been to improve the reliability that the maximum absolute stress will not violate the limit value of 620 MPa. Design variables in the optimization have been the mean values of the design parameters in Table 1. Assuming a constant relative standard deviation for each parameter, a reliability target of 4σ has been set. The targeted result of the RBDO is then to update the design parameters, such that the mean value of the maximum absolute stress will keep a distance of 4σ from the region of the event space where the maximum absolute stress has a value equal or smaller than 620 MPa. The reliability target of 4σ corresponds to a probability of $3.16 \cdot 10^{-5}$ that the maximum absolute stress over the three components will exceed 620 MPa.

The RBDO process has been carried out using the Performance Measure Approach and the HMV+ algorithm [5], looking for a robust optimum point in the design. The optimization resulted in a Reliability Index of 4σ, thus achieving the targeted lower probability of failure of $3.16 \cdot 10^{-5}$.

It is important to note that the time needed for the reliability analysis and the optimization process has been very limited, thanks to the use of the response surface model. In fact each FE computation of the structural and durability analysis takes a total of 45 minutes (mean time), while the response surface takes less than a second to evaluate. This makes it very clear that the Monte Carlo simulation and the optimization process would have been much more expensive, in terms of computational effort and time, without the use of a response model.

3.3 Controller design for virtual shaker. In the field of vibration testing, the interaction between the structure being tested and the instrumentation hardware used to perform the test is key. Especially for massive structures (e.g. satellites), the dynamics of the testing facility often couples with the test specimen dynamics in the frequency range of interest. The use of a closed loop real-time vibration control scheme could lead to pole shifts and change damping of the coupled system. "*Virtual shaker testing*" is a new approach to deal with these issues. A simulation is performed that closely represents the real vibration test setup, taking into account all parameters that may impact the specimen dynamics. In [33], a virtual shaker testing approach has been developed, consisting of a coupled electro-mechanical model and a vibration controller. The shaker table and controller must be designed to allow testing critical launch conditions without damaging the satellite. The controller performance must be guaranteed for the realistic range of satellite model uncertainty. A parameterized simulation model of the satellite in controlled virtual shaker conditions has been created, comprising a dynamic MBS model of the satellite and a controller model in Simulink [33]. The sine controller has been coupled to the satellite model, with the controller output directly applied as a force input to the structure (i.e. assuming a perfect shaker).

Since dynamic interaction is a critical issue in virtual shaker testing, it is key to assess the impact of input parameter uncertainty on the vibration modes and coupling thereof. In this section, an uncertainty assessment strategy has been adopted, using a fuzzy analysis with the TM, to assess the impact of uncertainty in the tested structure (modal parameters) on the controller performance. First, a parameter sensitivity study has been performed to assess the influence of uncertain satellite model parameters on the control quality. While mass and stiffness properties are relatively well known in general, understanding the nature and being able to quantify damping mechanisms in structures is a much more difficult task. Therefore, it is important to take large uncertainties into account when modeling damping. As case study, the two main satellite modes (i.e. the modes at 37 Hz and 42 Hz, which have the largest effective mass) have been selected as uncertain parameters. The spring and mass properties are kept constant, with a large variation of the damping ratio: 0.5 – 10 %.

The sine control test specification comes typically in the form of an acceleration spectrum that needs to be reproduced at the shaker table. In this case, the profile was linearly ramping up (on a log-log scale) between 1 and 10 Hz and kept constant till 100 Hz. After modeling the uncertainty in the modal damping characteristics, the Reduced TM has been used to assess the effect of uncertainties on the controller performance, for the design range of controller settings. Two controller parameters, the compression factor and the number of periods in the sine sweep, can be varied by the engineer to optimize the controller performance. It is of high interest to be able to determine for which controller parameters the impact of the (given) satellite uncertainty is minimal. A systematic uncertainty assessment has been performed for this purpose. A 3-level full factorial DOE has been performed on the two most important controller parameters (see definitions below). This allows showing the fuzzy uncertainty on the controller response in a 3-by-3 display of the fuzzy controller performance, see Figure 6.

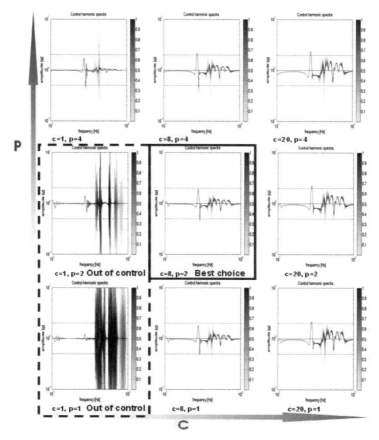

Figure 6: Virtual Shaker Testing: Uncertainty assessment of the controller performance, for constant model uncertainty, as function of the compression factor (c, horizontal) and the number of sine periods per estimation (p, vertical).

From Figure 6, one can identify both the worst-case scenarios and the best controller settings:
- Left to right, the compression factor (c), which determines the control agility, increases. A low factor means more agile control at the expense of noisier spectra and increased beating.
- Bottom to top, the number of sine periods (p) per estimation increases. Less period yields slightly better control, but less stable and noisier amplitude estimates.
- With a compression factor 1 (left column of Figure 6), the control is unstable. But increasing the number of periods improves the situation. The best choice in this study is the middle plot (c=8, p=2), with a controller performance that respects the alarm lines.

This case study offers a new perspective for the fuzzy FE method: it allows identifying the most robust controller settings for a given range of uncertainty of the controlled structural model.

Conclusions

This paper deals with the role and importance of uncertainties in different engineering design and test disciplines: crash and safety simulations, mechatronics system simulations and modal analysis. In real life systems, non-determinism is inherent component of the system model parameters, load conditions and the operational environment. Consequently, non-deterministic scatter characterizes any system's functional performance, which has important impact in the optimization design process, design for robustness and system identification procedures. In this paper, an outline has been given of methods to incorporate non-determinism (uncertainty and variability) into the virtual design process. Subsequently, the relevance of non-deterministic analysis in mechanical and mechatronics engineering has been demonstrated on a number of engineering application cases taken from automotive and aerospace industry. As such, the paper is aimed to contribute to raising the awareness of the added value of non-deterministic analysis, which has – up to this date – not yet become a mainstream approach in industrial product design. For a virtual shaker case study, the robustness of the controller performance (for a given test structure with constant uncertainty) has been assessed as a function of the controller settings (assessed through design of experiments). This new perspective for the fuzzy FE method has allowed identifying the most robust controller settings for the given range of uncertainty in the test structure.

Acknowledgements

We gratefully acknowledge IWT Vlaanderen for supporting the SBO project "Fuzzy Finite Element Method" and the O&O projects 070401 "I-CRASH", 080067 "MODELISAR" and 090408 "CHASING". Finally, we gratefully acknowledge the European Commission for supporting the FP7 Marie Curie IAPP project "TIRE-DYN" and the FP7 Marie Curie ITN "VECOM".

References

[1] T.A. Zang et al., Needs and Opportunities for Uncertainty-based Multidisciplinary Design Methods for Aerospace Vehicles, NASA/TM-2002-211462, 2002.

[2] W. Oberkampf et al., Variability, Uncertainty and Error in Computational Simulation, ASME-HTD-Vol.357-2, Proceedings of AIAA/ASME, pp. 259-272, 1998.

[3] R.B. Melchers, Structural Reliability Analysis and Prediction, 2nd Edition, 1999.

[4] G.I. Schuëller, H.J. Pradlwarter and P.S. Koutsourelakis, A Critical appraisal of reliability estimation procedures for high dimensions, Prob. Eng. Mech., Vol. 19, pp. 463-474, 2004.

[5] B.D. Youn, K.K. Choi and Y.H. Park, Hybrid Analysis Method for Reliability-Based Design Optimization, Journal of Mechanical Design, ASME, Vol. 125, pp 221-232, 2003.

[6] L.A. Zadeh, Fuzzy sets. Information and Control, 8:338-353, 1965.

[7] W. Dong and H.C. Shah, Vertex Method for Computing Functions of Fuzzy Variables, Fuzzy Sets and Systems 24, pp. 65-78, 1987.

[8] D. Moens and M. Hanss, Non-probabilistic finite element analysis for parametric uncertainty treatment in applied mechanics: Recent advances. FINEL 47(1), pp. 4-16, 2011.

[9] T.J. Lorenzen and V.L. Anderson, Design of Experiments, a no-name approach, Marcel Dekker, Inc., New York, USA, 1993.

[10] A.I. Khuri and J.A. Cornell, Response Surfaces, Design and Analysis, Marcel Dekker, Inc., New York, USA, second edition, 1996.

[11] M. Hanss, The Transformation Method for the simulation and analysis of systems with uncertain parameters, Fuzzy Sets and Systems 130, pp. 277-289, 2002.

[12] S. Donders, D. Vandepitte, J. Van de Peer, W. Desmet, Assessment of Uncertainty on Structural Dynamic Responses with the Short TM, J. Sound Vib 288 (3), pp. 523-549, 2005.

[13] S. Rao and J. Sawyer, Fuzzy finite element approach for the analysis of imprecisely defined systems, AIAA Journal, vol. 33, no. 12, pp. 2364–2370, 1995.

[14] B. Möller, W. Graf, and M. Beer, Fuzzy structural analysis using α-level optimization, Computational Mechanics, vol. 26, pp. 547–565, 2000.

[15] D. Degrauwe, Uncertainty propagation in structural analysis by fuzzy numbers. PhD thesis, K.U.Leuven, 2007.

[16] U. Akpan, T. Koko, I. Orisamolu, and B. Gallant, Fuzzy finite element analysis of smart structures, Smart Materials and Structures, vol. 10, pp. 273–284, 2001.

[17] D. Moens and D. Vandepitte, An interval finite element approach for the calculation of envelope frequency response functions, Int J Numer Meth Eng 61(14), pp. 2480-2507, 2004.

[18] L. Farkas, D. Moens and D. Vandepitte, "Fuzzy finite element analysis based on reanalysis techniques," Structural Safety, vol. 32, pp. 442–448, November 2010.

[19] M. De Munck, D. Moens, W. Desmet and D. Vandepitte, "An efficient response surface based optimisation method for non-deterministic harmonic and transient dynamic analysis," Computer Modeling in Engineering & Sciences, vol. 47, no. 2, pp. 119–166, 2009.

[20] M. Bulla and J.M. Terrier, Robust Crash Analysis, Proc Automotive CAE Grand Challenge 2010, Hanau, Germany, 30-31 March, 2010.

[21] R. Brown, CAE Robustness using a Crash Dummy Example, Proc Automotive CAE Grand Challenge 2010, Hanau, Germany, 30-31 March, 2010.

[22] P. Du Bois et al., Vehicle Crashworthiness and Occupant Protection, AISI, 2004.

[23] Aberdeen Group, "System Design: New Product Development for Mechatronics", 2008.

[24] Information on Safety Recalls, NHTSA, http://www-odi.nhtsa.dot.gov/recalls/, 2011.

[25] H. Van der Auweraer, B. Peeters and S. Donders, Importance of Uncertainty in Identifying and Using Modal Models, Proc. INCE Symposium, Le Mans, France, June 27-29, 2005.

[26] S. De Bruyne, H. Van der Auweraer and J. Anthonis, Advanced State Estimator Design for an Active Suspension, SIAT-2011-266, Proc. SIAT 2011, Pune, India, Jan. 19-22, 2011.

[27] W. Heylen, S. Lammens and P. Sas, Modal Analysis Theory and Testing, Dept. of Mech. Eng., K.U.Leuven, Belgium, 2004.

[28] B. Peeters, H. Van der Auweraer, P. Guillaume and J. Leuridan, The PolyMAX frequency-domain method: a new standard for modal parameter estimation?, Shock and Vibration 11, pp. 395-409, 2004.

[29] L. Soria, A. delle Carri, B. Peeters, J. Anthonis and H. Van der Auweraer, An operational modal analysis approach for the performance assessment of passenger car active suspension systems, Proc. ISMA 2010, pp. 595-625, Leuven (B), Sept. 20-22, 2010.

[30] L. Farkas, S. Donders, D. Schildermans, D. Moens and D. Vandepitte, Optimisation study of a vehicle bumper subsystem with fuzzy parameters, in Proceedings of International Conference on Uncertainty in Structural Dynamics, ISMA2010-USD2010, 2010.

[31] Allianz Zentrum für Technik GmbH (AZT), Geschäftsbereich Kraftfahrzeugtechnik, AZT Crashreparaturtest Front (Neuer RCAR Strukturtest - 10°, February 1, 2004.

[32] E. Isaksson, Simulation Methods for Bumper System Development, Licentiate Thesis, Luleå University of Technology, 2006.

[33] S. Ricci, B. Peeters, J. Debille, L. Britte and E. Faignet, Virtual shaker testing: a novel approach for improving vibration test performance, Proc. ISMA 2008, pp. 1767-1782, Leuven (B), Sept. 15-17, 2008.

[34] A. Friedmann, M. Schmidt, G. Rocca, J. Heimel, H. Buff and B. Peeters, Automated retrieval of modal properties for the monitoring of vehicle dampers, Proc. ICEDyn 2011, Tavira (P), June. 20-22, 2011.

[35] Noesis Solutions N.V., OPTIMUS Rev. 5.3 SP2, 2008.

[36] LMS International, "LMS Virtual.Lab", Rev 9, 2009.

[37] M. Hack, R. d'Ippolito, N. El Masri, S. Donders, N. Tzannetakis, Reliability based design optimization of the fatigue behaviour of a suspension system based on external loads. In K. Berns, C. Schindler, K. Dreßler, B. Jörg, R. Kalmar, & J. Hirth (Eds.), 1st Commercial Vehicle Technology Symposium (CVT 2010), pp. 277-286, Shaker Verlag, 2010.

Uncertainties with respect to active vibration control

Prof. Dr.-Ing Peter F. Pelz[1,a], Dipl.-Ing. Thomas Bedarff[2,b], Dipl.-Ing Johannes Mathias[3,c]

[1]Technische Universität Darmstadt, Chair of Fluid Systems Technology, Magdalenenstraße 4, 64289 Darmstadt, Germany

[2] Technische Universität Darmstadt, Chair of Fluid Systems Technology, Magdalenenstraße 4, 64289 Darmstadt, Germany

[3] Technische Universität Darmstadt, Fachgebiet Produktentwicklung und Maschinenelemente, Magdalenenstraße 4, 64289 Darmstadt, Germany

[a]peter.pelz@fst.tu-darmstadt.de, [b]thomas.bedarff@fst.tu-darmstadt.de, [c]mathias@pmd.tu-darmstadt.de

Keywords: active suspension system, vibration control, robust design

Abstract. The content of this work is the presentation of the prototype of a new active suspension system with an active air spring. As being part of the Collaborative Research Unit SFB805 "Control of Uncertainties in Load-Carrying Structures in Mechanical Engineering", founded by the Deutsche Forschungsgemeinschaft DFG, the presented active air suspension strut is the first result of the attempt to implement the following requirements to an active suspension system:

- Harshness and wear: Reduced coulomb friction, i.e. no dynamic seal.
- Plug and drive solution: Connected to the electrical power infrastructure of the vehicle.
- Vehicle and customer application by software and not by hardware adaption.

These requirements were defined at the very beginning of the project to address uncertainties in the life cycle of the product and the market needs.

The basic concept of the active air spring is the dynamic alteration of the so-called effective area. This effective area is the load carrying area A of a roller bellow and defined by $A := F/(p-p_a)$. F denotes the resulting force of the strut, p the absolute gas pressure and p_a the ambient pressure. The alteration of this effective area is realized by a mechanical power transmission, from a rotational movement to four radial translated piston segments. Due to the radial movement of the piston segments, the effective area A increases and so does finally the axial compression force F.

The prototype presented in this paper serves as a demonstrator to proof the concept of the shiftable piston segments. This prototype is designed to gather information about the static and dynamic behavior of the roller bellows. Measurements show the feasibility of the concept and the interrelationship between the piston diameter and the resulting spring force.

Introduction

Full Active suspension systems allow controlling heave, roll, and pitch motion of a vehicle body. In the following two different systems are compared. From the strengths and weaknesses analysis of those solutions the demands of innovative active suspensions are discussed.

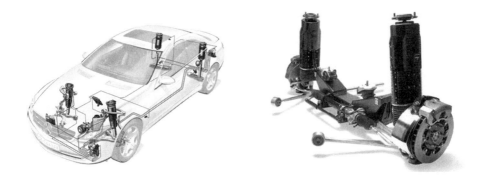

Fig 1: Left hand side, the Daimler active body control (ABC) [source: Daimler], right hand side, Bose-Suspension-System [source: Bose]

First (i) the up to now most advanced application of such an active suspension system for passenger cars is the active body control (ABC) suspension system invented by Daimler. The body motion is controlled up to a frequency of 5 Hz by means of hydraulically controlled servo cylinders in the four spring struts. The active base displacement of the coil spring together with the passive hydraulic shock controls the body movement [1]. Second (ii) for the electromagnetic Bose system [2], being still in the concept and not production phase, the base displacement of the torsional spring is realized by the electromagnetic force of a linear motor.

Tab. 1: Comparison of the Daimler and the Bose active suspension system. The points 1) to 4) are addressed in the text in more detail.

	hydraulic base displacement of a coil spring **(Daimler)**	electromagnetic base displacement of a torsion bar **(Bose)**
electrical power supply at each suspension strut	−	+
mass specific power	+	-- 1)
comfort at high frequency, small amplitude (Coulomb friction)	-- 2)	+
fail save (drivable without energy supply)	+	−
logistic and assembly costs for different suspension system on one vehicle platform	-- 3)	--
maintenance costs	- 3)	−
one hardware solution meets different OEM specifications by software adaption.	- 4)	−

Table 1 compares the hydraulic and the electromagnetic active suspension systems of Daimler, shown in Fig 1 left hand side, and Bose, shown in Fig. 1 right hand side. Even though such a comparison is always subjective and always critical, it should be done at the start of a new development. The four points especially marked in table 1 are discussed in more detail in the following:

1) *Power specific weight:*
   Comparing electromagnetic with hydrostatic motors, the mass specific power of the hydrostatic side is always favorable due to the high fluid pressure. This is the case for active suspension systems as well, where the power is determined by the body mass, the velocity and demand on control. Roughly the weight of electromagnetic devices is by one order of magnitude above the weight of the comparable hydrostatic device.

2) *Harshness due to coulomb friction:*
   As pointed out, the Daimler ABC system consists of a conventional hydraulic shock absorber and a (stiff) coil spring. Thus the suspension is harsh first due to the high eigenfrequency and second due to the Coulomb friction within the hydraulic shock absorber. For amplitudes smaller than the friction force divided by a typical suspension stiffness the harsh appearance increases significantly.

3) *Uncertainties due to Manufacturing and maintenance costs:*
   The logistic and assembly costs are significant in the case where for example the customer has the choice between an air suspension and an active body control. The complete chassis infrastructure depends on the customer choice.

4) *Uncertainties due to market needs:*
   Related to that point is the demand to change the vehicle dynamics not by changing the hardware, i.e. the suspension strut, but by changing the controller, i.e. by a software solution. Thus one suspension strut would meet the expectation of a BMW driver, who is looking for a sportive chassis, or a Mercedes driver, enjoying the driving comfort, and uncertainties in the market needs can be addressed.

Following the above realized analysis, three tasks were defined at the very beginning of the project to address uncertainties in the life cycle of the product and the market needs:

(i) *Harshness and wear:* Reduced coulomb friction, i.e. no dynamic seal.
(ii) *Plug and drive solution:* Connected to the electrical power infrastructure of the vehicle.
(iii) *Vehicle and customer application by software and not by hardware adaption.*

As being part of the Collaborative Research Unit SFB805 "Control of Uncertainties in Load-Carrying Structures in Mechanical Engineering", founded by the Deutsche Forschungsgemeinschaft (German research foundation), the main focus of our research was the question "how can we manage uncertainties in active suspension systems?".

Tab. 2: How to meet uncertainties in active suspension system

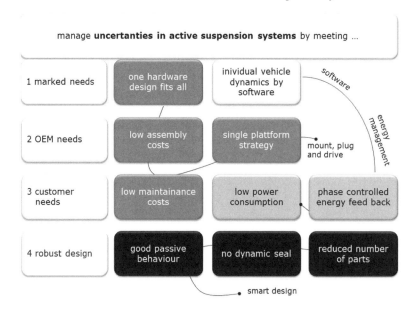

Table 2 gives a first reflection on that question. There is a clear difference between the overall market needs, the individual car manufacturer (OEM), and the later vehicle owner. Reviewing table 1 and 2 the design should be smart, it should be a plug and drive solution and the dynamic characteristics as well as the power management including recuperation - or as we call it, phase controlled energy feedback -, should be controllable by software.

Basic Concept

Air suspension is the state of the art suspension for luxury vehicles, sedan or SUV. The main reason for the success of air suspension systems is the invention of a thin (1.6 mm wall thickness) rolling bellow, which came to production in the Daimler S-Class in 1998. Essential for the success of an invention is the relation between customer values (which can also be emotional, i.e. subjective values) to costs for customer [3]. This ratio is superior for air spring systems in comparison to other suspension systems.

Fig. 2 shows two principle air spring designs in a schematic drawing.

The double roller bellows principle (Fig. 2b) allows a significant reduction of the load carrying area and is a promising concept to reduce coulomb friction, i.e. to reduce harshness [4].

The load carrying area A of an air spring bellows, sketched in Fig. 2a, is given by that diameter, where the bellows loop has a radial tangent. For the purpose of our research a very interesting concept is the double roller bellows concept shown in Fig. 2b. For this concept the load carrying area is given by $A=A_1-A_2$, where the index 1 denotes the upper bellows and the index 2 the lower one.

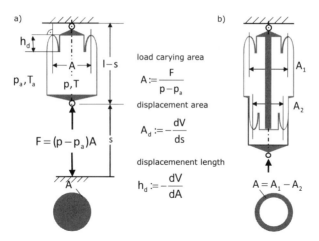

Fig. 2: Suspension strut including roller bellows: a) standard arrangement with one roller bellow and a rolling piston; b) two coaxial arranged roller bellows with the associated rolling pistons.

For all the fluid suspension systems shown in Fig. 3 the static equilibrium yields

$$F = (p - p_a)A, \tag{1}$$

where p denotes the absolute pressure (gas or liquid) within the device and p_a the ambient pressure. The very simple relation (1), which in turn serves to define the load carrying area, $A := F/(p-p_a)$ allows us to discuss all fluid suspension systems shown in Fig. 3 in a unified manner. Before doing so we have to define the displacement area A_d of the fluid suspension system as

$$A_d := -\frac{dV}{ds}, \tag{2}$$

where V is the gas volume and s the compression displacement of the suspension strut as shown in Fig. 2. The index "d" stands for displacement. For the special case of a plunger piston there is no difference in the area, $A_d \equiv A$, whereas for a rolling lope or bellows, A is slightly greater than A_d due to kinematic reasons. Similar we define a displacement length as

$$h_d := -\frac{dV}{dA}. \tag{3}$$

From the static equilibrium (1) the relative change in force follows as

$$\frac{dF}{F} = \frac{dA}{A} + \frac{dp}{p - p_a}. \tag{4}$$

Typical for a technical fluid suspension system the absolute pressure p is at least by one order of magnitude greater than the ambient pressure p_a. Hence in most cases it is justified to write

$$\frac{dF}{F} \approx \frac{dA}{A} + \frac{dp}{p}. \tag{5}$$

With the typical length of the suspension system defined by the gas volume divided by the typical area, which is the inverse of the volume specific area of the device,

$$l := V/A_d \approx V/A \tag{6}$$

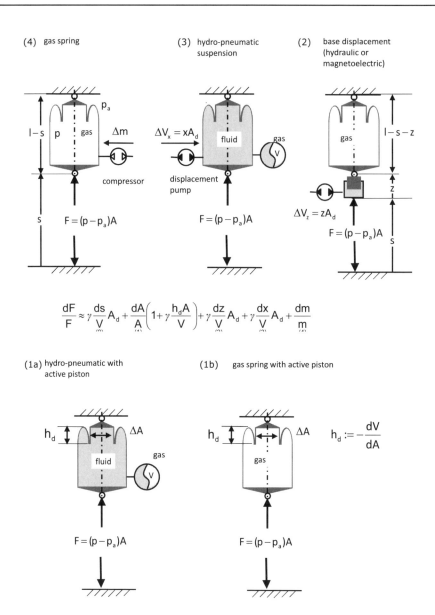

Fig. 3: Overview of fluid suspension systems: (4) air suspension; (3) active hydro-pneumatic; (2) active base displacement; (1a) and (1b) active change of the rolling piston (this research).

and the thermal diffusivity of the gas $a = \lambda / \rho c_p$, the cut-off-frequency for the transition from isothermal to isentropic change is given by

$$\omega_\gamma \sim \frac{a}{l^2}. \tag{7}$$

The heave eigenmode of an air suspended mass has the eigenfrequency

$$\omega_0 = \sqrt{\gamma \frac{g}{l}}. \tag{8}$$

Hence with $\omega_0 >> \omega_\gamma$ the condition for isotropic change of state is

$$\sqrt{\frac{a^2}{\gamma l^3 g}} << 1. \tag{9}$$

It is fulfilled for most technical fluid suspension systems for dynamic applications. Hence the isentropic relation

$$p(\rho) = const\, \rho^\gamma \tag{10}$$

is valid for the dynamic change of state, with the homogeneous gas density $\rho = m/V$. γ is the isentropic exponent which assumes the value of 7/5 for air.

With Eq. 10 the relative force change (Eq. 5) becomes

$$\frac{dF}{F} \approx \underset{(1)}{\frac{dA}{A}} - \gamma \underset{(0)...(3)}{\frac{dV}{V}} + \underset{(4)}{\frac{dm}{m}}. \tag{11}$$

The change of the gas mass dm within the suspension strut is in most cases quasi-static and hence isothermal, i.e. non isentropic. Thus the equation of state is isothermal and hence $p(\rho) = const\, \rho$.

The volume change in Eq. 11 can be considered to be

$$dV = -\underset{(0)}{A_d\, ds} - \underset{(1)}{h_d\, dA} - \underset{(2)}{A_d\, dz} - \underset{(3)}{A_d\, dx}. \tag{12}$$

The different terms in Eq. 11 and Eq. 12 are labeled by the very same labels used in Fig. 3. Thus the appearance of the different effects in different technical solutions for fluid suspension systems ranging from the classic air suspension system (term (0), (4)) hydro-pneumatic system (term (3)), base displacement system (term (2)), to the one we consider in our work, where we change the rolling piston area (term (1)), becomes obvious.

Thus the relative force change of the most general quasi-static fluid suspension system becomes

$$\frac{dF}{F} \approx \gamma \underset{(0)}{\frac{ds}{V} A_d} + \underset{(1)}{\frac{dA}{A}\left(1 + \gamma \frac{h_d A}{V}\right)} + \gamma \underset{(2)}{\frac{dz}{V} A_d} + \gamma \underset{(3)}{\frac{dx}{V} A_d} + \underset{(4)}{\frac{dm}{m}}. \tag{13}$$

As it becomes clear, Eq. 13 is valid for the conservative suspension strut (no damping force) in its equilibrium. Term (0) represents the spring force of the suspension system. Term (1) represents the relative force related to the change of the cross section area, which is addressed in our research. Term (3) represents an active base point displacement. As it becomes obvious this is from a physical point of view equivalent to an active change of the gas volume, which is done by active hydro-pneumatic solutions. The last term represents the change of gas mass which is achieved by a pneumatic infrastructure within the chassis (compressor, valves and air dryer).

Even though the terms (2) to (4) appear harmless, the infrastructure and hardware required within a suspension system to gain these effects are essential.

Due to the slowness of the gas supply system, the effect (4) can be used for quasi-static leveling purposes, but not for an active suspension system.

The advantages concentrating on the change in the carrying area are:

- There is no need for an external pump or compressor, i.e. the solution is a pump-less system.
- It is a "plug and drive" solution, since the internal actuator should be electrically driven.
- A small change in the cross section of the piston would result in a large change in the cross section area especially for the double bellow solution shown in Fig. 2 on the right.
- The package within the two pistons can be used to integrate the actuator.

Hence, for the mentioned reasons, it is worthwhile to consider only term (1) in Eq. 13 and to find a design solution for the principle given by

$$\frac{dF}{F} \approx \gamma \frac{ds}{V} A_d + \frac{dA}{A}\left(1+\gamma\frac{h_d}{l}\right), \tag{14}$$

for the one bellows solution and

$$\frac{dF}{F} \approx \gamma \frac{ds}{V} A_d + \frac{dA_1}{A}\left(C_1 - \frac{dA_2}{dA_1}C_2\right), \tag{15}$$

with the two constants $C_{1,2} = 1 \pm \gamma \frac{h_{d1,2}}{l}$.

for the double bellows solution. Preferably $dA_2/dA_1 < 0$ to enhance the actuator. The linearization of Eq. 15 yields

$$\frac{\Delta F}{mg} \approx \gamma \frac{s}{l} + \frac{\Delta A_1}{A}\left(1-\frac{dA_2}{dA_1}\right) \tag{16}$$

for $h_d << l = V/A_d$.

Transfer function of the active suspension system

The equation of motion for the supported mass reads (see Fig. 4 left hand side)

$$m\ddot{z} = \Delta F. \tag{17}$$

With Eq. 16 this results in

$$\ddot{z} \approx g\gamma\frac{s}{l} + g\frac{\Delta A_1}{A}\left(1-\frac{dA_2}{dA_1}\right) \tag{18}$$

with $s = z_0 - z$ or

$$\ddot{z} + \omega_0^2(l)z = \omega_0^2(l)z_0 + g\frac{\Delta A_1}{A}\left(1-\frac{dA_2}{dA_1}\right). \tag{19}$$

with the eigenfrequency $\omega_0 = \sqrt{\gamma\frac{g}{l}}$

Assuming a harmonic base displacement $z_0 = \hat{z}_0 \sin\Omega t$, the transmission function becomes

$$\frac{\hat{z}_+}{\hat{z}_0} = \frac{1 + \frac{l\Delta \hat{A}_1}{|\hat{z}_0|\gamma}\left(1-\frac{dA_2}{dA_1}\right)}{1-(\omega_0/\Omega)^2}. \tag{20}$$

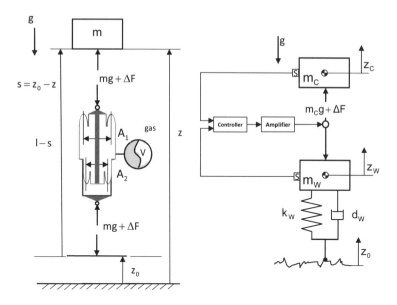

Fig 4: Spring-damper-mass system including the active suspension system (left); two degrees of freedom vehicle model used for the closed loop control (right).

The factor $(1-dA_2/dA_1)$ in Eq. 16 serves as an amplification factor. With the double bellows concept, the two pistons are coupled cinematically. Whenever the upper piston gets expanded $(dA_1>0)$, the lower one gets contracted $(dA_2<0)$ and vice versa.

Closed Loop Control

Fig. 4 on the right hand side shows the two degrees of freedom vehicle model, consisting of a cassis mass m_C, a wheel mass m_W and the active fluid suspension system as an actuator. For the passive system, the actuator is replaced by suspension with the stiffness k_C and a damper coefficient d_C. The wheel excitation is given by a measured road profile, taken from a country road for a speed of approximately 70 km/h.

With respect to vertical vehicle dynamics, the closed loop control is intended to meet two requirements:

1) Comfort control: to have the ride as comfortable as possible the acceleration of the chassis should vanish, i.e. $\ddot{z}_C = 0$.

2) Driving safety: to transmit transfer forces to the street, a minimal wheel load fluctuation is required.

These two goals are in conflict with each other, more driving comfort for example leads to less driving safety. This conflict is illustrated in Fig. 6: The driving comfort is given by the vibration intensity (in this work: not weighted), the driving safety by the normalized wheel load fluctuation. In Fig. 6, the wheel load fluctuation is normalized with the wheel load fluctuation given for an ideal actuator, i.e. the chassis acceleration vanishes.

For a passive system a variation of the chassis suspension spring stiffness and damper coefficient leads to the conflict diagram depicted in Fig. 6. A higher damper coefficient d_C of the chassis damper results in a reduced wheel load fluctuation (increasing driving safety), a lower chassis suspension spring stiffness k_C results in a reduced chassis acceleration (more driving comfort). With the passive system, is not possible to decrease both the chassis acceleration and the wheel load fluctuation beyond the cross curve. A lower wheel load fluctuation will always results in higher chassis acceleration. The white points in this conflict diagram indicate that the maximum relative travel of the suspension is exceeded.

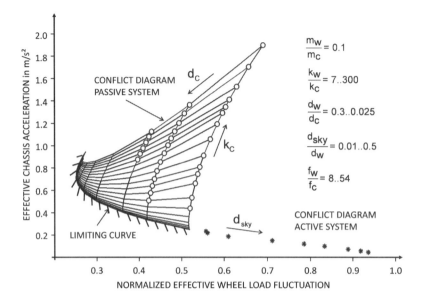

Fig. 6: Conflict diagram; comfort vs. driving safety

With an active suspension system, it is possible to overcome these restrictions. The goal of each active suspension system is to reduce the wheel load fluctuation as well as the chassis acceleration, i.e. reach the point of origin in the depicted conflict diagram. To do so, a sophisticated control algorithm is needed which deals with the vertical dynamics as well as with the longitudinal and lateral dynamics. In this work, only the vertical dynamics is considered. As a first approach, the active fluid suspension system is equipped with a combined sky-hook ground-hook controller. The sky-hook controller (damper coefficient d_{sky}) applies a force against the chassis velocity; the ground-hook controller applies a force against the wheel velocity. As expected, the chassis acceleration can be reduced significantly without violating the deflection limits. However, the wheel load fluctuation increases due to the concept of the controller.

Design concept of the new active fluid suspension system

Fig. 7 right hand side shows the above discussed change of the load carrying area in more detail. The piston is divided in segments which are forced radially outwards. Due to changes in the roller fold, the load carrying area changes as well. In Fig. 7 the load carrying area A_1 is enlarged and the load carrying area A_2 is reduced. Hence the difference cross sectional area $A=A_1-A_2$ is enlarged as well. Due to the fact that two pistons are used and A is significant smaller than A_1 or A_2, a small change of the single load carrying areas (A_1,A_2) will have a relatively large impact on the change of the difference cross sectional area A and hence a large impact on the relative force change $\Delta F/F$ given in Eq. 4.

Fig. 7 left hand side shows the principle effect of the alteration of the load carrying area as the result of a simple calculation: on the abscissae the diameter of the upper piston is plotted, on the left ordinate the resulting compression force at constant damper compression travel and on the right ordinate the associated diameter of the lower piston. The compression travel s of the suspension system is the parameter. In this graph, the labeled compression travel (-70 mm to +70 mm) is measured from the design position.

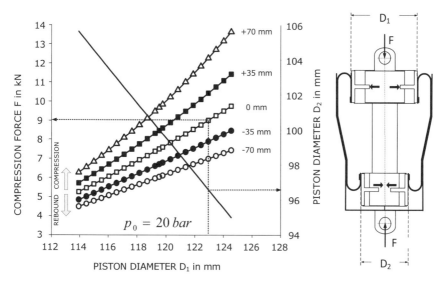

Fig. 7: Simulated compression force versus piston diameter for different suspension compression travel values s.

The white square markers in Fig. 7 show the change of the load carrying area in the design state. Within the project the design state is defined for a static compression force of 7.5 kN. The specification meets roughly the requirements of a luxury passenger vehicle.

Within the project for the final design solution of the active system, the lower piston always changes its diameter when the upper one does but in an opposite sense, to enhance the sensibility of the system. The dashed line in Fig. 7 shows: For a load of 9 kN the diameter D_1 is widened to 123 mm whereas D_2 is reduced to 96 mm. This spread in diameter increases with increasing compression travel s due to the increase in gas pressure.

Robust Design. For the technical realization of the alternating load carrying area a solution had to be found. The solution has to solve the following conflict: On the one hand there are large forces due to the pressure inside the bellows that must be overcome. And on the other hand the package space inside the piston is very limited. A feasible solution for this conflict is the radial shifting of the piston segments, described in the next section.

Fig. 8 shows the radial shifting of the piston segments. The thick line symbolizes the roller bellows. The problem with this kind of expansion is the interaction of bellows and piston. Throughout the detailed design of the piston segments two main uncertainties occur:

1. The problem seems to be the interaction of the bellows and the piston. Due to the pressure inside, the bellows adheres to the piston. Will the bellows glide along the piston surface as it is necessary to shift the segments?
2. An outward shifting of the piston segments, leads to a stretching of the bellows only between the segments. Will the bellows be destroyed by this stretching?

Based on design reviews with bellows experts the recommendation was to build up a test system to get more information about both uncertainties. In this test system bellows with different properties (rubber, fiber angle etc.) should be stretch inwards and outwards over many periods. Due to the results and the so increasing knowledge design changes for example of the bellows fiber angle should be discussed to design a working solution. It is easy to understand that this is a risk full approach for the designer. Design time, monetary and material effort will increase a lot. Additionally if the results are negative the designer needs a new solution or the whole concept may fail.

At this point it is very important for the designer to think about a design change that leads to a robust solution where the consideration of both uncertainties is not necessary anymore. By this approach no additional knowledge will be necessary saving time and money.

This robust solution should avoid a gliding between the bellows and the piston and reduce the stretching of the bellows to a minimum. A possible solution with both approaches integrated in a small but effective design change is shown in Fig. 8:

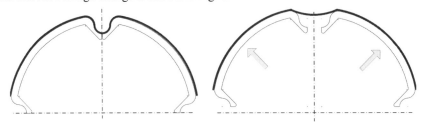

Fig. 8: Principle of the piston widening by shifting piston segments and by using a gap between the segments.

The robust solution is based on the special geometry of the piston surface. In Fig. 8 two piston segments are shown. A gap was put in between the two parts. The bellow lies in this gap when relaxed and gets tensioned when the piston segments shift outwards (the piston expands). With this technique the segments can be moved without putting too much strain on the bellow. No gilding of the bellow along the piston appears and the bellow is bended but not stretched. This bending has no problematic impact on the bellow as it is the main working principle of the rolling bellow.

Feasibility study of the concept by a finite element simulation. The practicability of this idea is tested by a numerical simulation. A nonlinear finite element model of the roller bellows was developed together with Vibracoustic GmbH & Co. KG, a company of the Freudenberg group, and consequently enhanced for this research. With the help of the numerical model the assembly process, starting from the installation of the roller bellows to the alteration of the load carrying area could be analyzed. The 1.6 mm thick roller bellows is modeled with solid continuum elements arranged in three layers: Two elastomer layers with a fiber reinforcement layer in between. The two fiber layer are laid to form a cross ply with a given angle between the fiber directions. This structure is modeled in continuum approach. The simulations, not shown here, show a robust solution not for all but some fiber materials.

Prototype and prove of concept. The technical realization of the shifting piston segments is shown in Fig. 9. An axle, similar to a camshaft is powered by a gear wheel (Fig. 9, left). The cam glides on hardened pads and pushes the piston segments outwards. The camshafts and the piston segments are mounted with floating bearings (linear bearings for the segments and radial bearings for the camshaft).

Four of these camshafts are mounted inside the piston and are powered by one gear wheel (Fig. 9 left hand side). The axle driving shaft is powered by a hydraulic swivel motor with a torque of up to 400 Nm at 200 bar hydraulic pressure. This high torque and the connected high power consumption are necessary because a force of up to several Kilo Newton is needed to move the segments. It is easy to explain, where these big forces come from: If for example the absolute pressure inside the bellows is 11 bar, an integration of this pressure over the area that is in contact with the piston (circumference 300 mm, height 35 mm) leads to a pressure related force of approximately 2.5 kN per segment. A solution for this challenge has to be developed in future work. The current concept mainly deals with the bellows expansion and helps to gather data and experience about the systems behavior.

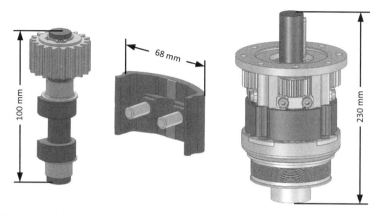

Fig. 9: Gearwheel driven camshaft (left), piston segment (middle) and complete assembled piston with four implemented piston segments.

In the Introduction the concept of two varying pistons was presented. The prototype though is a suspension strut with two pistons but only one of them, the top one, is variable. Therefore the influence of the segment shifting on the load carrying area is less, but still sufficient to show the principle feasibility and to collect information about the system behavior.

Fig. 10 shows the test-bench and a schematic diagram of it. The swivel motor is mounted on top of the upper piston and controlled with a closed loop circuit. The active suspension is mounted into an servo hydraulic test rig. Therewith it is possible to emboss the system with definite amplitudes and frequencies and to measure the resulting compression forces. The hydraulic swivel motor is powered by an axial piston pump (not shown in Fig. 10) which powers the camshaft. The following signals are measurement categories: The pressure in the two chambers of the swivel motor, the gas pressure, the temperature in the air spring, the amplitude and the speed of the basement excitation, the pivoting angle of the motor and the resulting spring force. Bases on the measured oil-pressure swivel motor the driving torque is calculated roughly. The measured pivoting angle serves to calculate the radial displacements of the piston segments.

The right side of Fig. 10 shows the principle structure of the prototype. For an easier assembly and to provide a linear guiding for the piston rod, the suspension strut is built with two roller bellows. The linear guide, also a plain bearing, prevents the air spring from buckling. The top piston is equipped with the moving segments, powered by the hydraulic swivel motor. The basement excitation is applied at the bottom part of the air spring, at the external guide.

The prototype has a weight of almost 60 kg and an overall height of 1100 mm. It is clear that this prototype will never be implemented in a car or truck. But this is not relevant because the intention for this prototype is, as already mentioned, to prove the presented concept in real life and to gain more information about the behavior of the roller bellows which is successfully done.

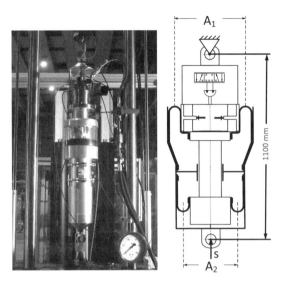

Fig. 10: Real test-bench (left hand side) and schematic diagram of the built active air spring (right hand side).

Fig. 11 depicts first results from a measurement, compared with a simple calculation based on the equations mentioned before. The abscissa shows the piston diameter, the ordinate the resulting force of the fluid suspension system. As expected, the resulting force increases during the variation of the load carrying area.

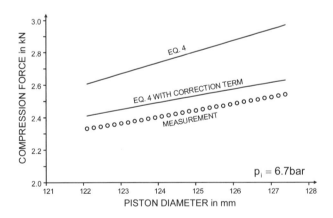

Fig. 11: Comparison of the measurements with the calculations

The calculation based on Eq. 4 assumes a circular load carrying area. In fact, the real load carrying area is smaller due to the gaps between the segments. The analysis of the nonlinear FE model results in a correction term, $(-0.007D_1+1.83)$, with the diameter of the upper Piston D_1. By multiplying the circular load carrying area with this area correction term (which results in the load carrying area given by the FE model), the measurement fits quite good the calculation.

Conclusion and Outlook

The first prototype of a new active fluid suspension system is introduced. The concept is a change of load carrying diameter of a hydro or pneumatic linear actuator. The presented prototype together with the discussed prove of concept is one major mail stone on the way to a final active suspension system. This one should not depend on an external hydraulic infrastructure (including pump, i.e. it is "pump-free solution"). It is free of Coulomb friction and hence robust in the sense of wear. Above that the stiffening at small amplitudes which is associated to Coulomb friction is reduced. By reducing the number of parts, the robustness of the system should be increases.

There were two design challenges to meet. First a concept of a segmented piston was presented. Second a mechanical transmission was designed to drive the piston segments.

Further investigations are needed to prove the capability of the presented concept and prototype for active vibration control.

List of Symbols

a	thermal diffusivity
A, A_d	load carrying area, displacement area
d_C, d_W, d_{sky}	damper coefficients of the chassis, the wheel and the sky-hook damper
f_c, f_w	chassis and wheel eigenfrequency
F	force
g	gravity constant
h_d	displacement length
k, k_C, k_W	stiffness, stiffness of the chassis suspension spring, stiffness of the wheel.
l	typical length
m, m_C, m_W	mass, chassis mass, wheel mass
p, p_a	pressure, ambient pressure
s	compression
V	volume
z	vertical coordinate
$z_0, \hat{z}_0$	base displacement, amplitude of the base displacement
γ	isentropic exponent
ω_γ, ω_0	cut off frequency, eigenfrequency
Ω	excitation frequency

References

[1] Pyper, M.; Schiffer, W. & Schneider, W. DaimlerChrysler (Ed.) ABC - Active Body Control, Verlag Moderne Industrie, 2003, 241

[2] Bose. Bose® Suspension System, Internet resource http://www.bose.com/pdf/technologies/bose_suspension_system.pdf, 2004

[3] Gies, S.: Bewertung geregelter Fahrwerksysteme aus der Sicht eines OEMs (engl: Evaluation of controlled suspension systems from the perspective of an OEM); Haus der Technik, Fahrwerktechnik 2.-4. Juni 2003, München

[4] Puff, M., Pelz, P. F., Mess: Influencing vehicle dynamics by Means of Controlled Air Spring Dampers; ATZ - Automobiltechnische Zeitschrift. Nr.: 2010-04

Effect of suspension parameter uncertainty on the dynamic behaviour of railway vehicles

Laura Mazzola[1], Stefano Bruni[1]

1 Politecnico di Milano, Dipartimento di Meccanica, Via La Masa 1 20156 Milano, Italy

laura.mazzola [stefano.bruni]@polimi.it

Keywords: railway design and performance verification, uncertainty in railway vehicle, Monte Carlo simulation

Abstract The paper describes a study carried out by Dipartimento di Meccanica Politecnico di Milano, aimed at investigating how uncertainty in railway vehicle suspension components can be treated in the framework of vehicle design and performance assessment in respect to vehicle dynamics. In railway vehicle suspensions, sources of parameter uncertainty may arise from inaccuracy in the modelling of a vehicle component or from a scatter in the behaviour of nominally identical components, on account of the variability implied by the component manufacturing process.

The approach proposed in this paper, completely new to the railway field, is to use statistical methods having different complexity (and entailing a proportional computational effort), to analyse the propagation of uncertainty from the parameters input in the vehicle mathematical model to the results of running dynamics, in terms of the assessment quantities used for verification and evaluation of train performances. The problem is treated by numerical means, being the dependency of simulation outputs from the input parameters typically non-linear, and not defined in an analytical form.

1. Introduction

The concept of reliability, probability or risk has been subject of extensive studies in many branches of engineering. In fact, any observable physical process contains a degree of uncertainty which shall be considered not only to evaluate the reliability, fitness of purpose and service life of systems and components, but more generally when approaching the design and engineering of any physical system. As far as mechanical systems are concerned, uncertainty effects have been mostly dealt with in the past with regard to durability issues, analysing in a probabilistic and risk-based perspective the ability of the component to survive the service loads for a given design lifetime. For the same systems, parameter uncertainty may also have serious implications on the capability to ensure the desired level of performance: this is a topic which has attracted until now limited research effort and deserves to be better investigated.

This paper deals with uncertainty effects in railway vehicles and their effect on the vehicle's running performances. More specifically, parameter uncertainty in the vehicle's suspension components is considered in view of its effect on the maximum admissible speed of the vehicle compatible with stability.

Two sources of parameter uncertainty are considered here: on the one hand, parameter uncertainty may arise from inaccuracy in the modelling of a vehicle component which, at least to some extent, can be represented in the mathematical model of the vehicle as a deviation of model parameters from their 'true' value. On the other hand, a scatter in the behaviour of nominally identical components can be expected, on account of the variability implied by the component manufacturing process. These effects cannot be quantified by means of traditional deterministic design approach on one simple train set, but can be handled by the integration of computer aided analysis and appropriate statistical tools [1, 2, 3].

The approach proposed in this paper, completely new to the railway field, is to use statistical methods having different complexity [4] (and entailing a proportional computational effort), to analyse the propagation of uncertainty from the parameters input in the vehicle mathematical model to the results of running dynamics. The problem is treated by numerical means, being the dependency of simulation outputs from the input parameters typically non-linear, and not defined in an analytical form.

In the paper the proposed numerical methodology for uncertainty analysis is presented, and applied to the estimate of the vehicle critical speed, demonstrating the possibility to include parameter uncertainty effects in the design of a railway vehicle in a simple and cost- effective way.

The paper is organised as follows: Section 2 describes the method used in this paper to estimate by means of multi-body simulation the critical speed of the vehicle, and introduces the parameters affecting the calculation and how uncertainty shall be considered on these quantities. Section 3 introduces a simplified linearised approach for quantifying the uncertainty of the railway vehicle critical speed, based on the assumed degree of uncertainty in the knowledge of vehicle suspension parameters, then in Section 4 more complex and comprehensive methods are introduced to analyse the propagation of uncertainty. In Section 5 some results of the two proposed approaches are analysed and compared, considering the case of a locomotive car from a concentrated power EMU train,. Finally, conclusions are drawn in Section 6.

2. Running dynamics evaluation: the uncertainty in the evaluation of vehicle service speed

Rail vehicles may suffer from a self-excited vibration mechanism known as "hunting" and consisting of the combination of a lateral and yaw oscillation of the bogies. This can be demonstrated to be in relation with the speed of the vehicle, so that the vehicle will be stable at low speed and unstable at higher speed. Therefore, one of the most important performance indexes which are carefully considered in the design stage of a new railway vehicle is the so called "critical speed" of the vehicle, i.e. the speed above which the "hunting instability" onsets. Clearly, the maximum admissible speed of the vehicle shall remain below the critical speed by an appropriate safety factor, so to guarantee that the vehicle can provide the required performance in terms of maximum service speed without undergoing dynamic problems which would seriously compromise the ride comfort, because of accelerated degradation of both the rolling stock and the infrastructure, and in serious cases might even affect the running safety.

In Europe, the critical speed of an existing vehicle can be experimentally determined according to the testing procedures reported in the EN14363 standard [5]. The same standard requires that the maximum vehicle service speed is 10% lower than the critical speed as determined from line tests.

More specifically, the Standard EN14363 defines two alternative criteria to determine experimentally the vehicle critical speed: the first one, referred to in the standard as the "simplified measuring method", is based on the measure of the bogie frame acceleration over the leading and trailing axles; these signals are band-pass filtered to select the frequencies associated with the hunting motion of the bogie and the standard deviation of the filtered signals $s\ddot{y}^+$ is defined, where the s indicates that the r.m.s. value of the signal need to be compared with the limit [5]. The critical speed of the vehicle is then defined as the minimum speed for which the assessment quantity $s\ddot{y}^+$ exceeds the limit value $\left(s\ddot{y}^+\right)_{\lim}$ defined as:

$$\left(s\ddot{y}^+\right)_{\lim} = \frac{1}{2}\left(12 - \frac{m_b}{5}\right) \qquad (1)$$

with m_b the bogie mass expressed in thousands of kilograms and being $\left(s\ddot{y}^+\right)_{\lim}$ expressed in m/s², while 12 is an acceleration value defined on basis of experience and expressed in in m/s² while factor 5 is expressed in kg/(m/s²).

According to the "normal measuring method" described by [5], the track shift forces (sum of the guiding forces on the two wheels of the same axle) are measured, and the standard deviation of the filtered contact force signal $s\Sigma Y$ is compared with the corresponding limit value $(s\Sigma Y)_{lim}$ defined as:

$$(s\Sigma Y)_{lim} = \frac{1}{2}\left(10 + \frac{2Q_0}{3}\right) \qquad (2)$$

where Q_0 is the static vertical wheel load in tons, and both Q_0 and $(s\Sigma Y)_{lim}$ are expressed in kN.

For the vehicle considered in this work (see section 5) the limit values for the assessment quantities are: $(s\ddot{y}^+)_{lim} = 4.8$ m/s² and $(s\Sigma Y)_{lim} = 30$ kN.

In the vehicle design stage, the critical speed has to be defined using multi-body models of the vehicle or of the complete train set, analysing the running behaviour of the system at increasing speeds. The classical way to deal with this problem is based on a linear stability analysis performed on the linearised equations of motion of the railway vehicle [6]. However, more recent investigations pointed out the importance of non-linear effects related with the wheel/rail contact and the actual geometry of wheel and rail profiles, and provided evidence for the onset of non-linear hunting limit cycles at speeds which can be considerably lower than the critical speed defined by the linearised approach [7, 8].

In the present study, a non-linear method to estimate the critical speed is adopted, based on the simulation of vehicle non-linear running behaviour in tangent track subject to the random excitation produced by track irregularity [9]. Numerical simulations are performed considering the straight track running of the vehicle in presence of random track irregularity, with stepwise increasing speed. For each vehicle speed considered, the standard deviation of both track shift forces and lateral bogie acceleration (pass-band filtered as explained above) are extracted, and the process is iterated until one of these two assessment quantities exceeds the corresponding limit as stated by the EN14363 Standard. The critical speed value is then obtained by linear interpolation between the two speed values falling across the limit condition.

In order to assess the effect of parameter uncertainty, the rail vehicle is considered as a mixed probabilistic-deterministic process, and the vehicle critical speed V is defined as a function of deterministic and random variables:

$$V = f(\underline{u}, \underline{p}) \qquad (3)$$

where $\underline{u} = \{u_1, u_2, ..., u_n\}^T$ is the vector of input variables affected by uncertainty, and $\underline{p}$ is the vector of deterministic variables in the problem, whose values are assumed to be known without uncertainty. Unfortunately no analytical expression is available for the dependency of the critical speed on the uncertain and deterministic parameters expressed by Eq. (3), hence the relationship defined by Eq. (1) can only be defined for discrete values of $\underline{u}$ and $\underline{p}$ using a multi-body model of the rail vehicle and the criteria established by EN14363 as outlined above.

Several suspension parameters can be considered to affect the vehicle critical speed and at the same time to be affected by uncertainty however, to confine the scope of the investigation, in this paper parametric uncertainty is assumed to occur only on the secondary yaw damper parameters (Fig. 2), defined as the viscous damping parameter c_d and the serial stiffness parameter k_d, according to the classical lumped model representation of yaw dampers [10]. Hence, the set of parameters affected by uncertainty is defined as follows:

$$\underline{u} = \{k_d, c_d\}^T \qquad (4)$$

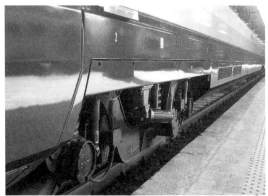
Figure 1: Side view of a train with yaw damper installed

The assumption is made that the parameters in vector $\underline{u}$ are statistically independent Gaussian variables, with the mean value corresponding to the nominal value of the parameter and standard deviation defined so that the 0.15-99.85 percentiles correspond, in the analysed case, to a ±15% variation of the yaw damper characteristic with respect to the nominal value, which is in line with the tolerances normally set on the supply of this railway component.

3. The "one-at-time" method to analyse of the effect of uncertain parameters

Different approaches can be adopted for the purpose of the present study. In this section a description of a simple one-at-time approach neglecting the interaction among the different uncertain parameters is introduced. Although neglecting the interactions may result into a low accuracy of the results of the uncertainty analysis, this method has the advantage of being very simple, allowing to consider the effect of uncertainty on a relatively large number of parameters at a low computational cost. It can be used therefore to obtain at least qualitative indications of the relative importance of uncertainty in different vehicle suspension parameters. From a quantitative point of view, the applicability of this method is discussed later on in the paper, when the results of the one-at-time approach are compared with those provided by a more comprehensive but also much more computationally demanding Monte-Carlo approach.

In the one-at-time approach a relationship between critical speed and input variability is defined based on the first order Taylor's expansion of equation (3) around the vehicle nominal condition, represented by the nominal value $\underline{u}_0$ of the uncertain parameters:

$$V = V_0(\underline{u}_0, \underline{p}) + \sum \alpha_i \cdot \Delta u_i \tag{5}$$

being V_0 the vehicle critical speed in nominal configuration and α_i the sensitivity coefficients expressing the linearised effect of parameter variations Δu_i on the critical speed. To compute the sensitivity coefficients, a one-at-time screening plan [11] of the mixed probabilistic/deterministic process defined by equation (3) is performed. For each uncertain parameter the nominal values and two extreme values are selected to define the effect of single parameters variation, and the sensitivity coefficients α_i are defined using the least square method.

Assuming a Gaussian probability density distribution for the uncertain parameters, and using the linearised input-output relationship, the probability density distribution obtained for the vehicle critical speed is also Gaussian, with mean μ_V and standard deviation σ_V defined according to fundamental statistics [1]

Using the above results, the vehicle critical speed can be defined in probabilistic terms, e.g. as the 0.15 percentile of the probability density distribution defined by μ_V and σ_V, which would mean a 99.85% probability of meeting the homologation requirement on the critical speed.

4. Monte Carlo and Design of Experiments based methods

In order to assess the effect of interaction between different uncertain parameters, a more comprehensive approach to the quantification of uncertainty effects in the calculation of the vehicle critical speed is introduced in this section, considering the simultaneous variation of all uncertain and assuming a non-linear dependence of the outputs from the uncertain input values [11].

In its simplest form, the non-linear approach foresees that each input parameter is sampled several times to represent its real distribution according to its probabilistic characteristic defined as described in section 2. Then combining all the input parameters, several sets of input values are obtained, each one representing one "realisation" of the input-output problem. Solving the input-output problem for each realization (in this case by means of numerical simulation), the overall probabilistic characteristics of the output is identified. Although the implementation of this approach is relatively simple, the numerical procedure entailed may become severely time consuming if a large amount of input parameters are considered.

The most straightforward implementation of this approach consists of a Monte Carlo simulation (MCS), see Figure 2: firstly a set of random parameter values are generated for inputs (i.e stiffness and damping characterising the bogie yaw damper) according to their assumed probabilistic distribution; Then for each realisation of the random variables set previously defined the corresponding output (i.e. vehicle critical speed) is computed performing non-linear simulation of the vehicle running behaviour. Finally a statistical analysis is carried out on the output sample (i.e the sample of critical speed values obtained as the result of the MCS) to identify the minimum possible critical speed for the vehicle configuration analysed.

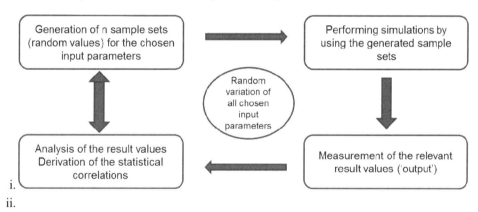

iii. Figure 2: Propagation of uncertainty: Monte Carlo Simulation techniques

The generation of random input parameters according to their assumed probabilistic distribution is the core of the Monte-Carlo simulation, since this will affect the convergence of the method, especially when the number of realisations is to be kept to a minimum to reduce the size of the experiment. A reasonable level of computational efficiency can be achieved through the appropriate choice of the sampling method [1].

Among the different sampling methods proposed by the variance reduction techniques the Latin Hypercube (LH) one has been chosen for application to the problem treated here. This method enables to better cover the domain of variations of the input parameters, thanks to a stratified sampling strategy, and can be applied in the case of independent input parameters, which is the case assumed here. The Latin Hypercube procedure is based on dividing the range of each parameter into several intervals of equal probability and undertaking the sampling as follows: first the range of each input parameter is stratified into iso-probabilistic cells, then a cell is uniformly chosen among all the available cells and the random number is obtained by inverting the Cumulative Density

Function locally in the chosen cell, finally all the cells having a common strate with the previous cell are put apart from the list of available cells [2]. This method enables the efficient exploration of the domain of variations of the input parameters. Additionally, one can show that the LH strategy leads to a variance reduction if the model is monotonic over each variable, but as previously pointed out it is valid only if the input random variables are independent.

As an alternative a second approach the Monte Carlo Simulation technique has been coupled with the Design of experiment theory allowing for a more efficient and less time consuming method. This second approach is defined as follows:

v. A full factorial plane is design to account for parameter interaction and non - linearity. [12]. the factors level are representative of the parameters range of variation;

vi. Non - linear simulations are performed for the combinations at point i), and the output (i.e. vehicle critical speed) for each combination is computed;

vii. The output of point ii) are used to defined a polynomial relationship between output deviation and the parameter variation. The validity of the relationship is then properly verified;

viii. Random parameter values are generated for the inputs according to their assumed probabilistic distribution;

ix. For each realisation of the random variables set defined at point iv) the corresponding output is computed on the basis of the relationship assessed at point iii);

x. For the output sample obtained a full statistical analysis is then carried out.

The reliability and accuracy of the results provided by this approach is highly dependent on the definition of the relationship between output variability and input uncertainty.

To correctly represent this connection, a 3-level full factorial plane is as adopted to account for non - linearity and interactions. In the designed plane, three values are selected for each uncertain input, corresponding to the extremes and central point of the input range of variation

Considering all different combinations in the values of the parameters under study, a total of 3^{n_p} "experiments" are defined, being n_p the number of parameters being varied.

For all these experiments, the vehicle critical speed is computed (cf. Section 2) and the dependence of this quantity on damper parameters is then approximated over the entire range of the yaw damper stiffness and damping parameter values using a polynomial expression of the type [12]:

xi.
$$\hat{V} = V_0 + a_1 \cdot x_1 + a_2 \cdot x_2 + a_3 \cdot x_1 \cdot x_2 + \\ + a_4 \cdot x_1^2 + a_5 \cdot x_2^2 + a_6 \cdot x_1 \cdot x_2^2 + a_7 \cdot x_1^2 \cdot x_2 + a_8 \cdot x_1^2 \cdot x_2^2$$
(6)

Where a_i are the coefficients of the relationship model, calculated applying the least square method to the data, x_1 the stiffness parameter k_d and x_2 the damping parameter c_d.

5 Propagation of uncertainty in running dynamics evaluation: results

In this section the results of propagation of uncertainty is analysed applying the three methodologies described above (i.e. 1. OAT, 2. MCS, 3. MCS & DOE). A concentrated power locomotive for high speed train with maximum service speed 240 km/h is considered and therefore, based on EN14363 prescription, a minimum critical speed of 264 km/h (110% of the maximum service speed) is required. The effect of uncertainty is analysed below in terms of probability of not meeting the above stated requisite on the critical speed.

The analyses performed showed that the leading bogie in the vehicle is more critical than the trailing one in terms of both the assessment quantities foreseen by the simplified and normal measuring methods, see Section 2. Therefore, all results presented below in this section are referring to the front bogie.

First, the application of the one-at-time method is shown. Three different values are considered for the stiffness and damping parameters, corresponding to the nominal one and to the statistical minima and maxima, for a total number of 5 vehicle configurations for which the critical speed value is defined by means of multi-body simulation using the method described in Section 2.

Table 1 reports the critical speed values obtained for different combination of the damper parameters c_d and k_d, the sensitivity coefficients expressing the linearised relationship between the damper parameters and the critical speed. Based on these results, the probability of not meeting the stability criterion introduced (critical speed above 264 km/h) is defined as:

xii. Table 1: One-at time method Effect of yaw damper parameter variations on the critical speed

Stiffness [kN/m]	9625	11323	13022	11323	11323
Damping [kNs/m]	345	345	345	293	397
Vehicle Critical speed [km/h]	264	271	278	266	274
Sensitivity coefficient	0.004 [(km/h)/kN/m]			0.080 [(km/h)/kNs/m]	
Standard deviation of the critical speed [km/h]	3,87				
Pf(Vcr<264 km/h)	3.5 %				

xiii.

xiv.

As an alternative approach, the Monte Carlo simulation technique is applied to the same case study. The Monte Carlo experiment implemented, accounts for 100 samples, corresponding to 100 simulation cases defined based on random combined variation of parameters c_d and k_d (Fig. 3a) This number of simulation cases is large but still affordable in terms of CPU time.

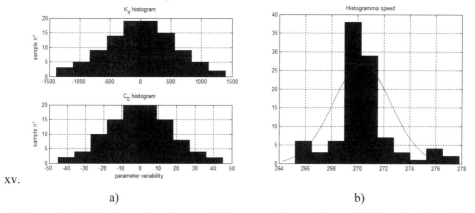

xv.

a)   b)

xvi. Figure 3: Distribution of random variable generated according to LHS method a) stiffness and damping samples distribution (100 samples) b) critical speed distribution (100 samples)

Unfortunately the chosen number of samples is demonstrated not to be sufficient for correctly reproducing the distribution of the critical speed (cf. Figure 3b), in fact only a small number of samples describe the tails of the distribution, hence the accuracy of the statistical analysis cannot be guaranteed. A significant higher number of simulations was expected to be needed in order to obtain reliable results. Therefore the application of the pure Monte Carlo simulation technique appeared to be not feasible in terms of time cost and hence then DOE & MCS method is implemented after a suitable validation of the methodology with respect to pure Monte Carlo simulation technique.

Based on the DOE theory a full factorial plane is defined and the relationship between the critical speed and the varied parameters. For the analysed case n_p =2, hence the total number of numerical experiments to be performed is 9 (section 4). Then the critical speed sample is realised with an increasing number of Monte Carlo simulations, using the polynomial function (6) to define the input-output. The statistical processing of the outputs was then carried out iteratively for increasing size of the Monte Carlo experiment, until convergence of the critical speed probability density function was reached for a sample size of 1000 units, see Figure 4. Finally, the probability of meeting the required critical speed of 264 km/h was computed for this output sample, see Table 2.

As demonstrated in Fig. 4 a and b the Gaussian distribution assumption appears to be valid also in correspondence of the tails, despite the non-linear input-output relationship defined according to equation (6).

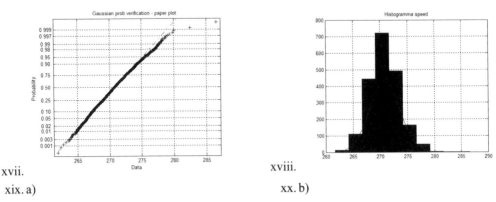

xvii.   xviii.
xix. a)   xx. b)

xxi. Figure 4: a) Paper plot for 1000 samples b) Critical speed histogram for 1000 samples

The DOE and MCS approach allows to identify a probability of meeting the critical speed of 0.6% much lower than the one computed with the one-at-time method.

The results of the one-at-time method approach and the DOE&MCS are compared in table 2, in terms of resulting mean value, standard deviation as well as 0.15 percentile and 99.85 percentile. A non-negligible difference is observed on the standard deviation and therefore on the tails of the two distributions. If the mean values are analysed it appears that they correspond almost the critical speed the vehicle has when in nominal configuration. Furthermore due to the higher standard deviation value, the probability of the critical speed exceeding the imposed thresholds has similar value, besides the OAT evaluation is conservative, being higher the probability of meeting the critical speed computed

Table 2: OAT and DOE&MCS method comparison

	OAT	DOE&MCS
Mean Critical speed [km/h]	271.0	270.8
Std Critical speed [km/h]	3.9	2.8
Pf (Vcr<264 km/h) [km/h]	3.5%	0.6%

6   Conclusions

This paper presented a study finalised to the investigation on how parametric uncertainty can be treated in the framework of railway vehicles design in respect to vehicle dynamics, considering uncertainty in the yaw damper properties and considering full correlation between parameters of all dampers in the vehicle. This type of uncertainty can be associated to inaccuracy in the modelling of a vehicle component which, at least to some extent, can be represented in the mathematical model of the vehicle as a deviation of model parameters from their 'true' value.

Three approaches with different level of complexity were proposed in the paper to analyse the propagation of uncertainty from the suspension component parameters to the critical speed of the vehicle. The results presented in the paper demonstrate that the suitability of the proposed numerical methods to account for parameter uncertainty effects in the estimation of the critical speed for a railway vehicle.

The results presented in the paper also show that simplified methods, such as the OAT one can be adopted when preliminary investigation are carried out, whereas the approach based on the combined use of Design of Experiments and Monte Carlo simulation shall be used, despite the larger computational effort required, when quantitative results are sought for.

Finally, the presented approaches allow to verify the train performances already during the design stage, thus offering the possibility to reduce the amount of line testing and the associated cost and effort. At the same time, these methods underpin a more objective homologation process, allowing the investigation of issues not addressed by the physical homologation process, such as the impact of uncertainties in components behavior.

7   References

[1] A.Haldar, S.Mahadevan 'Probability, reliability and statistical methods in engineering *design'*, John Wiley 2000

[2] J.C. Helton, F.J. Davis. "Latin Hypercube sampling and the propagation of uncertainty analyses of complex systems". *Reliability Engineering and System Safety 81*, p.23-69, 2003

[3] D.G. Cacuci Sensitivity and Uncertainty Analysis. *Vol. 1. Theory* (CRC,2003)(T)(285s) CHAPMAN & HALL/CRC

[4] J.S.Liu 'Monte Carlo stategies in scientific computing',*Springer 2001*

[5] EN 14363: *Railway Applications — Testing for the Acceptance of Running Characteristics of Railway Vehicles —* Testing of Running Behaviour and Stationary Tests, CEN, Brussels, June 2005

[6] A. Wickens Fundamentals of Rail Vehicle Dynamics – Guidance and Stability, 2003, Swets & Zeitlinger

[7] H. True, "On the Theory of Nonlinear Dynamics and its Applications in Vehicle Systems Dynamics", Vehicle System Dynamics, 31 (5), 393-421, 1999.

[8] O. Polach, "On non-linear methods of bogie stability assessment using computer simulations", Proc. Inst. Mech. Eng. F. J. Rail Rapid Transit, 220 (1), 13–27, 2006.

[9] S.Alfi, S.Bruni, L. Mazzola: "Effect of motor connection on the critical speed of high speed railway vehicles", *Vehicle Syst. Dyn, 46(Suppl), 201-214,2008*

[10] S. Bruni, J. Vinolas, M. Berg, O. Polach, S. Stichel, "Modelling of suspension components in a rail vehicle dynamics context", accepted for publication on Vehicle System Dynamics

[11] A., Saltelli, K., Chan, M. Scott, "Sensitivity Analysis", *John Wiley & Sons publishers, Probability and Statistics series, 2000*

[12] D.C.Montgomery.: Design and analysis of experiment, *John Wiley &Sons 2001*

Evaluation and control of uncertainty in using an active column system

Jan F. Koenen[1,a], Georg C. Enss[1,b], Serge Ondoua[1,c],
Roland Platz[2,d] and Holger Hanselka[1,2,e]

[1]Technische Universität Darmstadt, System Reliability and Machine Acoustics SzM,
Magdalenenstrasse 4, 64289 Darmstadt, Germany

[2]Fraunhofer Institute for Structural Durability and System Reliability LBF,
Bartningstrasse 47, 64289 Darmstadt, Germany

[a]koenen@szm.tu-darmstadt.de, [b]enss@szm.tu-darmstadt.de, [c]ondoua@szm.tu-darmstadt.de,
[d]roland.platz@lbf.fraunhofer.de, [e]hanselka@szm.tu-darmstadt.de

Keywords: uncertainty, active systems, load-monitoring, active stabilisation

Abstract Uncertainty in usage of load-carrying systems mainly results from not fully known loads and strength. This article discusses basic approaches to control uncertainty in usage of load-carrying systems by passive and active means. An active low damped column system critical to buckling is presented in which a slender column can be stabilised actively by piezo stack actuators at one of its ends only. Uncertainty may be controlled in the active column system by temporarily enhancing the bearable axial load theoretically up to three times compared to the passive column system in case of critical loading. However, in the implementation of these approaches, system-specific uncertainty may also occur. In numerical examinations it is shown, that small deviations in measured axial loading may increase the active force significantly to achieve stabilisation. The increase of applied active force might affect lifetime of the piezo stack actuators and thus the stabilising capability of the active column system.

Introduction and motivation

Uncertainty generally occurs in processes of the product life cycle of load-carrying systems, [1]. Especially in usage of load-carrying systems, knowledge of loads and strength of the structure play an important role to predict its lifetime, reliability etc. Design engineers of load-carrying systems always try to prove functional suitability during product development of the product by means of computational or experimental examination of the operational stability and reliability to, eventually, control uncertainty e.g. due to deviations in geometry, of material properties or of assumed loading etc. According to VDI 4001 [2], reliability describes the probability an item will perform its required function without failure under stated conditions for a stated period of time. Thus, to predict reliability of a load-carrying system, it is essential to know actual loads and strength. However, during usage sufficient knowledge of actual loads as well as existing strength is uncertain in most cases.

This study is part of the Collaborative Research Centre (SFB) 805: *Control of Uncertainty in Load-Carrying Systems in Mechanical Engineering*, which is publicly funded by the Deutsche Forschungsgemeinschaft (DFG). Within this workgroup, engineers and mathematicians have set up the following working hypothesis of uncertainty: "Uncertainty occurs when process properties of a system can not or only partially be determined", [1].

Today, expected loading conditions acting on load-carrying systems are often preliminarily estimated by simplified or reduced load spectra. Real loads are often estimated by standards or in real experiments in laboratory tests or field trials. For example, in some standards the maximal loads acting on truss structures are only estimated due to reasonable assumptions on the mass and on the maximal accelerations, e.g. [3]. Sometimes, standardised load spectra provide approximated information about height and frequency of loads for some groups of load-carrying systems, e.g. *TWIST*, a standardised load spectrum for bending moments in wing root areas of airplane, [4]. Field trials with prototypes may deliver product- and application-specific load spectra that are assumed to be representative for similar products, [4],[5]. On the other hand, strength of structures can be predicted by methods of computer aided engineering, e.g. finite element based methods. However, this procedure delivers only a cut-out of future usage, uncertainty remains in any prediction.

Fig. 1 describes the relationship between uncertainty in loads and strength and introduces a safety factor S.

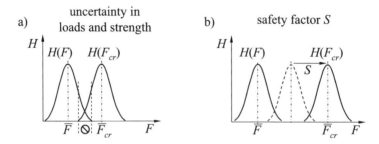

Fig. 1: Uncertainty in loads and strength: a) significant failure probability, b) standard solution: using safety factor S

Herein, $\overline{F}$ is the expected load value and $\overline{F}_{cr}$ is the expected strength value, whereby the index cr stands for critical if a strength value in the application should not be exceeded, e.g. tensile stresses or elongation of fracture etc. $H(F)$ and $H(F_{cr})$ are assumed frequency distributions of load and strength. They represent uncertainty and are assumed as normally distributed for simplicity. In reality, however, distributions may not even be known. In fig. 1a, the intersection of $H(F)$ and $H(F_{cr})$ represents a significant probability for F exceeding F_{cr} which may lead to a functional failure. Thus, uncertainty of not fully known real loading condition and strength often requires the use of safety factors S in product development to get a sufficient buffer between acual load and strength of the product (safe-life), fig. 1b. For example, safety factors $S = 3 - 10$ are provided for slender structures critical to bucking, [6]. Furthermore, a load-carrying structure is designed to resist the worst case load scenario even if limit loads are reached in few occasions. This normally leads to an oversizing of the developed system and may increase the deployed economical and ecological ressources.

Fig. 2 shows different approaches to control uncertainty in a general manner. For example, usage-monitoring provides the possibility to analyse actual loading and deformation conditions online individually for a structure so that the actual scatter of loads could be known more detailed over its lifetime, $H(F)$ becomes narrower, fig. 2a, [7]. If the actual loading and deformation condition is known, it may be possible to actively affect the deformation state by using active components and a suitable control algorithm to eventually increase the strength of the system, maybe only for a short but sufficient time, fig. 2b. Compared to high strength of the structure due to the use of safety factors, strength increase would be necessary only on demand when the actual loading condition is critical concerning a failure of the system. Thus, a load-carrying system could be designed to withstand most actual normal loads without being oversized, but also withstand rare cases with high and critical loads.

Using active components requires a particular focus on investigation of interactions between active and passive system components for instance in active system strength evaluation. As shown in fig. 2c, active system strength F_{cr} may be estimated so that its distribution function $H(F_{cr})$ shows a narrow variance. However, actual critical strength F_{cr} may show a higher variance over time caused by many factors like scatter in components properties, e.g. deviations in stiffness or capacity of piezo actuators. Additionally, the overall active system strength may decrease over time due to degradation of components. As a consequence, again load and strength distribution functions may intersect and an increase of probability of failures during usage could be possible, fig. 2d. Thus, the uncertainty in the strength of passive and active components should be identified and evaluated for the whole lifetime of active systems.

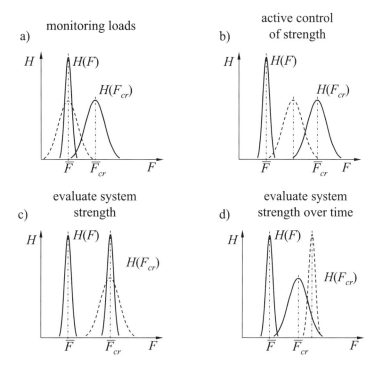

Fig. 2: Approaches to control uncertainty: a) monitoring loads, b) increasing strength by means of active control, c) evaluate strength of passive and active components d) evaluate strength over time

Technologies like usage-monitoring or active control and evaluation can therefore make an important contribution to control uncertainty in usage of load-carrying systems. However, within several processes of those technologies, uncertainty also may occur. Thus, if these technologies should be deployed, an important part will be to describe, to evaluate and finally to control those system-specific uncertainties. The following chapter gives an example by a simple load-carrying active column system.

Description of active column system

In this chapter, a simple active column system will be introduced to discuss approaches to reduce uncertainty by active control in using load-carrying systems, [8]. Thereby, this active column system is an example for any load-carrying active system, so that knowledge and methods gained in the project could be transferred to other load-carrying active systems. Fig. 3a and 3b show the schematic sketch of an uncontrolled and controlled column system, fig. 3c shows the real built system.

A flat slender column of length L with constant bending stiffness EI, constant mass distribution ρA and damping θ is loaded with an axial load P. It is pinned on the upper end $x = L$ and clamped at the lower end $x = 0$. If the axial load P at $x = L$ on the column reaches its first critical buckling load $P_{cr,1}$, (4), the straight column may buckle in its first buckling mode shape W_1 due to overload $P > P_{cr,1}$, fig. 3a.

Buckling control of the active column is based on the combination of two stabilising control concepts. First, the initial buckling deflection should be compensated through a counteracting active force F_a close to the column's base at $L_a \gg 0$ to cancel even smallest deflections that may lead to sudden buckling, fig. 3a. Second, the column could be enforced to vibrate in its second bending deflection

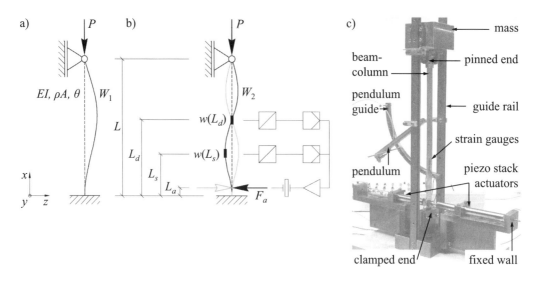

Fig. 3: Demonstrator: schematic sketch of a) uncontrolled and b) actively controlled column system, c) picture of real demonstrator

mode shape W_2, fig. 3b. This results in reduced buckling length L_d due to a steady state vibration node at $x = L_d$ and thus a critical load $P_{cr,2} \approx 3 \cdot P_{cr,1}$ which can be, theoretically, three times higher than the critical buckling load $P_{cr,1}$ of a straight column for the given boundary conditions, [8].

However, the column's stiffness and therefore also its eigenfrequencies ω_n decrease with increasing axial load P as explained in the following chapter. Thus, assurance of the axial load P is essential to excite the column in its second eigenfrequency ω_2, see following chapter. This can be done for example by continuously measuring P. The column's deflection $w(L_s)$ is chosen in numerical estimation to have a constant vibration amplitude sufficient to achieve stabilisation, [8]. The second control concept for stabilisation will be examined in this paper with regard to uncertainty.

The active system in fig. 3 consists of four main components: passive beam-column, strain gauge sensors, piezo stack actuators, fig. 3c, and information processing. Information processing contains the measuring chain from sensor signal to actuator signal with sensor and actuator amplifiers, cables, lowpass filters, A/D-converter, realtime control system, D/A-converter, etc. Further components of the real demonstrator are a pendulum limited by a pendulum guide to induce disturbance forces of defined magnitude in later experiments, guide rails to support the mass which applies the axial load P on the beam-column which may buckle only in one plane and a fixed wall to support the piezo stack actuators, fig. 3c. During usage of the active system, uncertainties may occur in using all four main components. For example, the actual loading on the passive structure may be unknown or stiffness of the clamped end may change during usage. If the column's strain is measured at L_s to monitor the system's deflection state, uncertainties may occur in the measurement e.g. due to measurement noise or systematic deviations, so the measured signal does not match the real state. This may have an effect on the identified axial load P and thus on the stabilising potential, i.e. load-carrying capacity $P_{cr,2}$ of the active column system, [8]. Uncertainties in information processing might also occur e.g. due to nonlinear behaviour as voltage or frequency limitations of the chosen piezo amplifier or discretisation errors in A/D-conversion. Furthermore, uncertainties occur in the actuators, e.g. scatter in the electromechanical properties of piezoelectric actuators that may lead to scatter in the active force F_a. Tab. represents the properties of the real active column system.

In this article, two possible approaches are discussed to control or, respectively, reduce uncertainty in an actively controlled load-carrying system critical to buckling, fig. 3. This is achieved by engaging

Tab. 1: Properties of the built active column system

Geometric properties	Material properties	Boundary conditions / Mech. properties
$L = 300$ mm	Aluminium: AlMg3	pinned end at $x = L$
$B = 20$ mm	$E = 70$ kN/mm^2	clamped base at $x = 0$
$H = 1$ mm	$\rho = 2660$ kg/m^3	P_l in N, constant at $x = L$, x-direction
$L_d = 168$ mm	$\theta = 0.01$	Sensor positions at $x = L_s$ and L_d, z-direction
$L_s = 100$ mm		F_a in N, active force at $x = L_a$, z-direction
$L_a = 10$ mm		

a usage monitoring system to identify the axial load P and by active stabilisation control. In the following, first a mathematical description of the active column system is presented to discuss influence of identifying the axial load P for activating the force F_a.

Example: Effect of uncertainty in measurement on the active force

In this chapter, a mathematical description for the dynamical behaviour of the column system as shown in fig. 3 is presented. With this, the influence of uncertainty in measuring axial load P on controlling the active force F_a, needed to achieve stabilisation, will be discussed. Thereby, only the second bending mode shape is considered, which is relevant for active stabilisation.

To achieve stabilisation of the column, a vibration of the second bending eigenmode with an amplitude $w(L_s)$ is needed. This is realised by a harmonic active force with an amplitude

$$F_a(\Omega) = |C_{dyn}(\Omega)| \cdot w(L_s) \tag{1}$$

and with the complex dynamic stiffness

$$C_{dyn}(\Omega) = m_{g,2} \cdot \frac{-\Omega^2 + 2\theta_2 i\Omega\omega_2(P) + \omega_2^2(P)}{W_2(L_a)W_2(L_s)} \tag{2}$$

in frequency domain, wherein Ω is the excitation frequency, ω_2 is the second eigenfrequeny of the column, θ_2 the modal damping and $m_{g,2}$ the modal mass for the second mode. $W_2(L_a)$ and $W_2(L_s)$ are the deflections of the second bending mode shape W_2 at $x = L_a$ for the active force F_a and at $x = L_s$ for the sensor, see fig. 3. However, according to [9], the second eigenfrequency ω_2 is a function of the axial load P, the second critical buckling load $P_{cr,2}$ and the second eigenfrequency ω_2^0 for the column with axial load $P = 0$,

$$\omega_2(P) = \omega_2^0 \sqrt{1 - \frac{P}{P_{cr,2}}}. \tag{3}$$

Generally, the n-th critical buckling load leads to

$$P_{cr,n} = \lambda_n^2 EI, \quad \text{whereby} \quad \tan\lambda_n - \lambda_n = 0 \tag{4}$$

with the n-th eigenvalue λ_n according to Timoshenko's theory of elastic stability, [10]. Fig. 4 shows the relation described in (2) and (3) for the second mode, $n = 2$.

If the structure is excited in its second eigenfrequency $\Omega = \omega_2$, the dynamic stiffness for the second mode has a minimum, so that a minimal acive force amplitude according to (1)

$$F_a(\omega_2) = \frac{m_{g,2}w(L_s)}{W_2(L_a)W_2(L_s)} \cdot |-\omega_2^2 + 2\theta_2 i\omega_2^2 + \omega_2^2|$$

$$= \frac{m_{g,2}w(L_s)}{W_2(L_a)W_2(L_s)} \cdot |2\theta_2 i\omega_2^2| \tag{5}$$

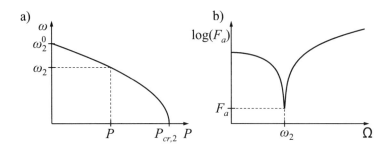

Fig. 4: a) Second eigenfrequency w_2 depending on axial load P; b) active force F_a depending on excitation frequency Ω. A minimal active force is needed, when $\Omega = w_2$.

is needed, fig. 4b. Thus, to minimise the active force and with this minimising the loading and energy consumption of the active system, the aim is to excite the structure at or near its second eigenfrequency w_2. Therefore, the axial load P needs to be measured and (3) yields to $w_2(P)$.

Now, one important aim to control uncertainty in load-carrying systems is to understand propagation of uncertainty in processes like usage processes. As an example, the effect of a small unknown deviation of the axial load $\Delta P = \tilde{P} - P$ between the affecting actual axial load P and the measured, uncertainty afflicted axial load $\tilde{P}$ on controlling the active force F_a will be discussed in the following. The tilde indicates an uncertainty afflicted value.

In this paper, it is assumed for uncertainty investigation that the column is loaded with an axial load P smaller than its second buckling load $P_{cr,2}$ but higher than its first buckling load $P_{cr,1}$. The active column system as described previously can carry this load by active stabilisation due to enforcing the column in its second eigenform W_2. As explained in fig. 4b and according to (5), a minimal active force F_a is needed to ensure a vibration amplitude $w(L_s)$ that is sufficient to induce the second mode and, therefore, stabilisation. Due to a deviation Δw_2, however, the structure will not be excited in its real second eigenfrequency w_2, but in uncertainty afflicted frequency $\tilde{w}_2$. Thereby, ΔP appears, for example due to uncertain calibration of the sensor or a unconsidered temperature influence. As it is shown in the following, unknown deviation ΔP may lead to an increase ΔF_a of the active force.

Eq. (3) describes the relation between the axial load P, the critical buckling load $P_{cr,2}$ and the second eigenfrequency $w_2(P)$. A deviation ΔP in the measured axial force leads to a deviation Δw_2 of the assumed second eigenfrequency,

$$\Delta w_2 = \tilde{w}_2(\tilde{P}) - w_2(P)$$
$$= w_2^0 \left(\sqrt{1 - \frac{\tilde{P}}{P_{cr,2}}} - \sqrt{1 - \frac{P}{P_{cr,2}}} \right) \tag{6}$$

by a control algorithm. Thus, a deviation Δw_2 will lead to an increase ΔF_a of the excitation forces

$$\Delta F_a = \tilde{F}_a(\tilde{w}_2) - F_a(w_2)$$
$$= \frac{m_{g,2} w(L_s)}{W_2(L_a) W_2(L_s)} \cdot \left(|-\tilde{w}_2^2 + 2\theta_2 i w_2 \tilde{w}_2 + w_2^2)| - |(2\theta_2 i w_2^2)| \right), \tag{7}$$

according to (1) and (2). Fig. 5 shows the relation described in (6) and (7). Due to small modal damping θ_2, ΔF_a is sensitive to a deviation Δw_2.

Tab. and fig. 6 show results from numerical simulation for three different assumed cases A) $P/P_{cr,2} = 0.5$, B) $P/P_{cr,2} = 0.7$ and C) $P/P_{cr,2} = 0.9$ of deviating axial loads P acting on the

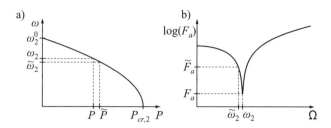

Fig. 5: Relation between a deviation $\tilde{P}/P$ in measured axial load and resulting inrease $\tilde{F}_a/F_a$ of excitation force amplitude

column. Numerical values of geometrical and material properties of the column system are taken from tab. and fig. 3. The corresponding second eigenfreqency for the column with axial load $P = 0$ is $\omega_2^0 = 822.1$ rad/s and the corresponding second critical buckling load according to (4) is $P_{cr,2} = 77.40$ N. A vibration amplitude $w(L_s) = 2$ mm in the second mode is assumed to achieve stabilisation. A deviation between the actual axial load P and the uncertain axial load $\tilde{P}$ is assumed to be $\pm 5\%$.

Tab. 2: Effect of $\tilde{P}/P$ relation on the excitation forces $\tilde{F}_a/F_a$ relation

Case	$P/P_{cr,2}$	P [N]	$\tilde{P}$ [N]	$\tilde{P}/P$	ω_2 [rad/s]	$\tilde{\omega}_2$ [rad/s]	$\tilde{\omega}_2/\omega_2$	F_a	$\tilde{F}_a$	$\tilde{F}_a/F_a$
A	0.5	38.70	40.63	1.05	581.3	566.6	0.97	3.35	8.99	2.68
			36.77	0.95		595.6	1.02		9.05	2.70
B	0.7	54.18	56.89	1.05	450.2	423.2	0.94	2.01	11.87	5.90
			51.47	0.95		475.8	1.06		11.91	5.93
C	0.9	69.66	73.14	1.05	260.0	192.8	0.74	0.67	15.08	22.51
			66.18	0.95		313.0	1.20		15.09	22.53

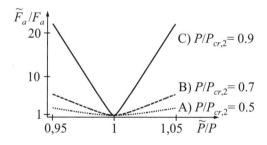

Fig. 6: Relation between relative deviation $0.95 \leq \tilde{P}/P \leq 1.05$ in measured axial load and resulting relative inrease $\tilde{F}_a/F_a$ of excitation force amplitude for three different cases:
A) $P/P_{cr,2} = 0.5$, B) $P/P_{cr,2} = 0.7$ and C) $P/P_{cr,2} = 0.9$

In tab. and fig. 6 it can be seen, that with increasing absolute value of deviation $|\tilde{P}/P|$, the active force $\tilde{F}_a$ increases significantly, if P becomes almost critical with $P/P_{cr,2} = 0.9$. This effect, in turn, increases with increasing axial load P even more. For example, with an axial load of $P/P_{cr,2} = 0.9$ and a deviation of $\pm 5\%$ of uncertain axial load $\tilde{P}$, the active force $\tilde{F}_a$ increases with a factor of 22.5 to achieve stabilisation. This effect mainly results due to low damping of the column and thus the increase of dynamical stiffness with deviation between real defined eigenfrequency ω_2 and uncertainty afflicted frequency $\tilde{\omega}_2$, fig. 5.

Conclusion and outlook

Uncertainty in the use of load-carrying systems mainly results from not fully known loads and strength. This article discusses four basic approaches to control uncertainty in use of load-carrying systems by implementation of technologies using active components: 1. monitoring loads, 2. increasing strength by means of active control, 3. evaluating strength of passive and active components and 4. evaluating active systems strength over time. An active column system was presented to discuss the influence of uncertainty in measured axial loading on the active force. It was shown, that a deviation of ±5% in the measured axial loading increases the active force up to a factor of 22.5 to achieve stabilisation in this simple example of a beam column sensitive to buckling. The simple example shows, that the functionality of active load-carrying systems may be higly sensitive to uncertainty. Further aims should be to describe uncertainty in active load-carrying system as comprehensively as possible and make use of existing methods and technologies to reduce and finally control uncertainty in a holistic approach.

To estimate the expected lifetime of the active system, the relation between the force amplitude and the bearable load cycles must be known quantitatively, e.g. due to a S/N-curve. Until now, research on fatique of piezoceramics is still in its beginning, which is an uncertainty itself, [11]. To handle this uncertainty, increasing the amount of available information due to extensive measurement or experiments on the lifetime of a specific piezo stack actuator may help to control this uncertainty.

Experiments of the presented numerical work are being prepared and experimental results will be presented soon.

Acknowledgements

The authors would like to thank the Deutsche Forschungsgemeinschaft (DFG) for funding this project within the Collaborative Research Centre (SFB) 805.

References

[1] H. Hanselka, R. Platz: *Ansätze und Maßnahmen zur Beherrschung von Unsicherheit in lasttragenden Systemen des Maschinenbaus,* "Controlling uncertainties in load-carrying systems", Konstruktion, 6, pp. 55 to 62, Springer VDI-Verlag, (2010).

[2] VDI 4001 Standard, *General guide to the VDI-handbook reliability engineering* VDI-standard, Beuth Verlag, Berlin, Wien, Zürich, (1985)-10.

[3] DIN15018: Krane, *Grundsätze für Stahltragwerke, Berechnung,* "Principles for steel structures, calculation"', (1984).

[4] E. Haibach; *Betriebsfestigkeit: Verfahren und Daten zur Bauteilberechnung,* "Structural durability: Methods and data for components calculation", 3rd edition, Springer, (2006).

[5] M. Buerle-Mahler; *Lastdatenaufnahme und Ermüdungsfestigkeits- und Lebensdauervorhersage,* "Load data recording and fatique strength and durability prediction", Books on Demand, (2008).

[6] K.-H. Grote and W. Beitz (Ed.): *Dubbel: Taschenbuch für den Maschinenbau,* "Dubbel: Handbook for mechanical engineering", Springer, (2001).

[7] D. Söffker; *Zur Online-Bestimmung von Zuverlässigkeits- und Nutzungskenngrössen innerhalb des SRCE-Konzeptes,* "Online-Determination of Reliability Characteristics as a Modul", Automatisierungstechnik, 8, pp. 383 to 391, (2000).

[8] G.C. Enss, R. Platz, H. Hanselka: *A survey on uncertainty on the control of an active column critical to buckling,* ICEDyn, 17, (2011).

[9] P. Hagedorn, S. Otterbein: *Technische Schwingungslehre 2, "Mechanical vibrations"*, Springer Berlin Heidelberg New York, (1987).

[10] S.P. Timoshenko, J.M. Gere: *Theory of Elastic Stability*, McGraw-Hill Book Company Inc., (1961).

[11] J. Nuffer: *Ermüdung piezokeramischer Aktorwerkstoffe -- Forschungsbedarf auf Bauteil- und Systemebene, "Fatigue of piezoceramic actuator material -- research needs on component- and systemlevel"*, Konstruktion 4, Springer VDI-Verlag, (2005).

Influence of the Selected Fatigue Characteristics of the Material on Calculated Fatigue Life under Variable Amplitude Loading

NIESŁONY Adam[1,a], KUREK Andrzej[1,b]

[1]Opole University of Technology, Faculty of Mechanical Engineering,
Department of Mechanics and Machine Design, ul. Mikołajczyka 5, 45-271 Opole, POLAND

[a] a.nieslony@po.opole.pl, [b] a.kurek@doktorant.po.edu.pl

Keywords: variable amplitude loading; cumulative damage; stress-life; strain-life; fatigue life assessment

Abstract. The algorithm of fatigue life determination for machine elements subjected to random loading uses fatigue characteristics of the material determined under constant-amplitude loading. They are usually stress or strain characteristics as well as characteristics using the energy parameter. Their correct selection influences correctness of the obtained results related to the experimental data. The paper presents analysis of convergence of the calculated fatigue lives of some constructional materials subjected to random loading under uniaxial loading state. For calculations concerning one material the same loading state was assumed and fatigue characteristics were determined on the basis of one data set obtained under constant strain amplitude tests. Calculated fatigue lives based on different fatigue characteristics were compared and their convergences were tested. It has been proved that convergences are different depending on the material. The comparison results were presented in form of graphs.

Introduction

Analysis of the fatigue life of elements of modern machines and advanced structures is a necessary element of a design process. At present, high requirements should be met by the final product and the design offices have better and better calculation potential, so calculations are more and more often performed on the assumption that loading are variable-amplitude or random [1-3]. Usually the calculations are performed at two basic stages while fatigue life determination under uniaxial random loading. The first is to define the load and the choice of size, which will be qualitatively and quantitatively describe the load. Engineers usually apply the stress history or the strain history, the energy parameter is rarely determined. At the second stage, the fatigue damage degree is calculated for the assumed loading block. In this case, a suitable hypothesis of fatigue damage summation is used. This task can be realized if distribution of the loading cycle amplitudes for the random history is known. Such distribution is obtained according to special algorithms of cycle counting, for example the rain flow algorithm [4], or one of the standard loadings/distributions typical for a certain group of devices and environments is assumed (for example FALSTAF distribution, applied in aviation, or CARLOS, used in car industry).

Depending on the quantity chosen for loading description, the calculation algorithm includes a suitable fatigue characteristic of the material, defining a number of cycles to the fatigue fracture versus the chosen parameter. For example if the stress amplitude is chosen then the Basquin equation can be used for description of materials fatigue [1, 2, 5]

$$\sigma_a = \sigma'_f (2N_f)^b, \tag{1}$$

where σ_a is the stress amplitude, N_f is a number of cycles up to fracture, σ'_f and b are the coefficient and the exponent of fatigue strength, respectively. The Manson-Coffin-Basquin (MCB) strain characteristic [5-8] is another popular fatigue characteristic of material

$$\varepsilon_{a,t} = \varepsilon_{a,e} + \varepsilon_{a,p} = \frac{\sigma'_f}{E}(2N_f)^b + \varepsilon'_f(2N_f)^c, \qquad (2)$$

where $\varepsilon_{a,t}$ is the total strain amplitude expressed by the sum of elastic and plastic strain amplitudes ($\varepsilon_{a,e}$ and $\varepsilon_{a,p}$, respectively), E is Young's modulus, ε'_f and c are the coefficient and exponent of plastic strain, respectively. The MCB characteristic is based on the controversial assumption that the description of the plastic and elastic parts of the total strain amplitude by straight lines in the double-logarithmic system is correct. In many papers we can find the data proving that this assumption is not correct for some materials [8, 9]. In particular, it concerns the term responsible for description of the plastic strain amplitude $\varepsilon_{a,p}$. This phenomenon observed for AISI 304 steel is shown in Fig. 1. Thus, there are different strain models joining the total strain amplitude with the number of cycles up to the fatigue fracture, for example the Manson proposal [6]

$$\varepsilon_{a,t} = \left(\frac{C}{N_f}\right)^{\frac{1}{\xi}} + \varepsilon_0, \qquad (3)$$

where C, ξ and ε_0 are the model coefficients determined by fitting the curve to the experimental points with the least square method.

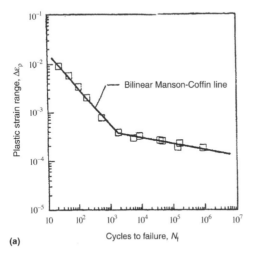
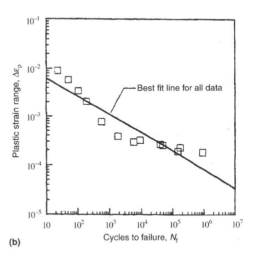

Fig. 1. Fatigue curves ($\varepsilon_{a,p}$-N_f) in the double-logarithmic system for unalloyed steel AISI 304: proposal of approximation of experimental points by two intervals better shows real changes (a) and approximation by one straight line according to the MCB characteristic (b), see [8] for details

Besides stress and strain also some energy parameters are used for description the materials fatigue. Because there are different definitions of those parameters a great number of such characteristics exist in the literature [10, 12]. They are usually obtained by transformations of formulas resulting from the stress-life and strain-life models. Let us consider an exemplary characteristic obtained by comparison of the Smith-Watson-Topper parameter (SWT) $\varepsilon_a\sigma_{max}$ with the product of the Basquin stress characteristic (1) and the strain MCB characteristic (2) [12]

$$\varepsilon_{a,t}\sigma_{max} = \frac{(\sigma'_f)^2}{E}(2N_f)^{2b} + \sigma'_f\varepsilon'_f(2N_f)^{c+b}. \qquad (4)$$

All the discussed fatigue characteristics expressed in Eqs. (1)-(4) can be drawn on the basis of the results obtained from one test series. They are typical tests described in the standards [13], realized under the constant strain amplitude $\varepsilon_{a,t}$ with registration of a number of cycles up to failure N_f and the stress amplitude σ_a. Only one pair of stress and strain amplitudes corresponds to one number of cycles, so it is expected that fatigue life determination with use of different characteristics should lead to the same calculated fatigue lives.

In this paper the authors compared the fatigue lives calculated from the mentioned fatigue characteristics, assuming the same state of random loading for each calculation path. Simulation calculations were performed for four selected materials, four types of amplitude distributions and fatigue characteristics according to Basquin (1), MCB (2), Manson (3) and SWT (4). The problems connected with multiaxial loadings and selections of the hypothesis of fatigue damage summation were not considered because the authors did not want to introduce additional factors influencing the results of calculations.

Algorithm of fatigue life determination

While determining the fatigue life under random loading, the component amplitudes are identified in the considered loading history. Thus, the history is replaced by a certain finite number of loading cycles with the sinusoidal course and the known amplitudes. This procedure is shown in Fig. 2.

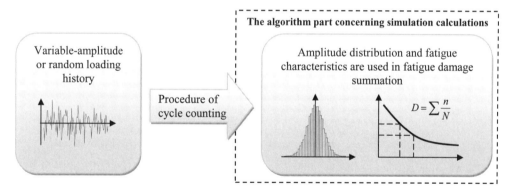

Fig. 2. Replacement of the random history by amplitude distribution in the algorithm of fatigue life determination under random loading

While further calculations the determined cycles are treated similarly like short constant-amplitude loading intervals. According to the linear hypothesis concerning fatigue damage summation, the damage is expressed by the following expression

$$D = \sum_{i=1}^{k} \frac{n_i}{N_i}, \qquad (5)$$

where n_i is a number of cycles with a known amplitude, N_i is a number of cycles up to failure read out from the suitable fatigue characteristic for the i-th loading amplitude.

Fig. 3 presents a general scheme of the algorithm applied during simulation calculations concerning fatigue life determination according to various fatigue characteristics. Four materials were subjected to simulation calculations: three steels – Ck45, X8CrNiTi810, 15Mo3, and one aluminium alloy – AlMg5.1Mn. In literature there were full data concerning fatigue tests under amplitude-constant loading [14]. The basic material constants are presented in Table 1. The material constants occurring in the Basquin model (1), MCB (2) and SWT (4) were taken from literature [14]. The coefficients occurring in the Manson model were obtained by fitting the curve defined by Eq. (3) to the experimental points ($\varepsilon_{a,t,i}$, N_{fi}) with the least square method.

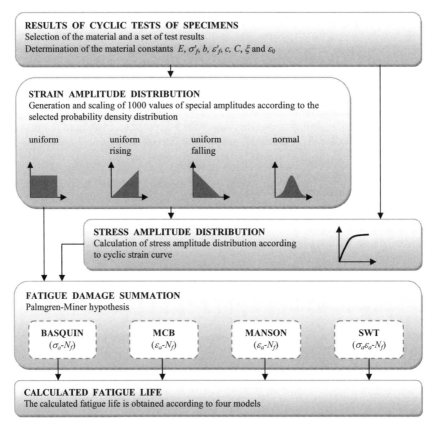

Fig. 3. Algorithm of fatigue life determination according to different characteristics of the material used during simulation calculations

In order to limit a number of the algorithm operations influencing the resulting calculated fatigue life it was assumed that the total strain amplitudes were generated directly from four selected probability density functions described by distributions included into the algorithm (see Fig.3).

During simulation calculations, 1000 values of the total strain amplitudes were generated. Before generation, the distributions were scaled so as their boundaries corresponded to the strain amplitude values from the MCB curve for the determined limit numbers of cycles $N_{f\min} = 10^3$ and $N_{f\max} = 10^6$. In such a way, the amplitude distribution including so-called low- and high-cycle fatigue ranges was obtained for each considered material. MCB characteristics and boundaries are shown on Fig. 4.

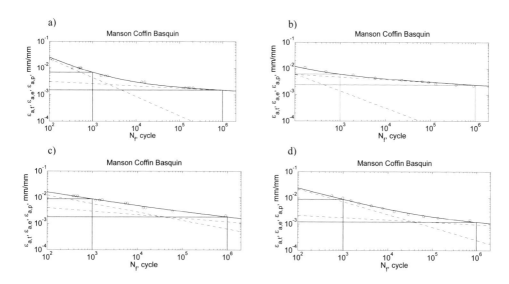

Fig. 4. Manson-Coffin-Basquin curves for selected materials, showing ε_a range used for generation: a) CK45, b) AlMg5,1Mn, c) X8CrNiTi1810, d) 15Mo3.

Table 1. Material constants of four considered materials

Material constant	Material			
	CK45	AlMg5,1Mn	X8CrNiTi1810	15Mo3
E, [GPa]	206	69.05	204	211
K', [MPa]	980	636.4	5234	893
n'	0.115	0.0843	0.421	0.18
σ'_f [MPa]	987	701.94	1658	706
b	−0.0828	−0,1014	−0.1343	−0.0870
ε'_f	0.9936	0.1955	0,0669	0.293
c	−0.7147	−0.6505	−0.3226	−0.4897
ξ	1.884	2.765	2.75	1.691
C	0.0581	0.0005527	0.0014	0.3007
ε_0	0.0014	0.0017	0.0014	0.0013

In the case of the Basquin stress model (1) and the SWT model (4), the stress amplitudes σ_a should be known. Suitable values were calculated from the cyclic hardening curve described with the Ramberg-Osgood model (RO) [15]

$$\varepsilon_{a,t} = \varepsilon_{a,e} + \varepsilon_{a,p} = \frac{\sigma_a}{E} + \left(\frac{\sigma_a}{K'}\right)^{\frac{1}{n'}}, \qquad (6)$$

on the assumption that the mean value of stress cycles was equal to zero [16]. The RO model is nonlinear, so the probability density function for stress amplitudes differs from the assumed probability density function for the strain amplitudes from which it was determined.

The determined pairs of values (ε_a, σ_a) participate in the fatigue damage summation. The Palmgren-Miner linear hypothesis was chosen, and the damage value $D = 1$ was defined, so it was possible to calculate the fatigue life expressed in loading cycles for the considered case

$$N_{obl} = \frac{1000}{D_{blk}}, \tag{7}$$

where D_{blk} is the damage value for one loading block, and the number 1000 corresponds to the number of the generated cycles participating in the fatigue damage summation. Two from four formulas describing the chosen fatigue characteristics allow for a short notation of the general formula (5). For the Basquin model (1) we obtain

$$D_{blk} = \sum_{i=1}^{k} \frac{2n_i}{\left(\frac{\sigma_{a,i}}{\sigma'_f}\right)^{\frac{1}{b}}}, \tag{8}$$

and for the Manson model (3) we have

$$D_{blk} = \sum_{i=1}^{k} \frac{n_i(\varepsilon_{a,t,i} - \varepsilon_0)^{\frac{1}{c}}}{C}. \tag{9}$$

Determination of a number of cycles N_i for strain amplitudes $\varepsilon_{a,t,i}$ and stress amplitudes $\sigma_{a,i}$ from the formulas MCB (2) and SWT (4) was realized with a numerical method. The determined numbers of cycles were directly introduced into Eq. (5).

Conclusions and final remarks

The calculated fatigue lives are presented as bar charts in Fig. 5. It is easy to find that generation according to the uniform falling distribution causes a visible increase of the fatigue life. It is caused by a low number of amplitudes of values close to the maximum one in the distribution, causing the greatest damage. The remaining fatigue lives were a little different in the decreasing order for normal, uniform and uniform rising distributions.

The calculations according to the Manson characteristic show its different nature as compared with the other characteristics. It can be a result of the fact that it was formulated according to different theoretical assumptions (no division into elastic and plastic parts of the strain amplitudes) and it has the strain fatigue limit ε_0.

The most interesting is that for the aluminium alloy calculations showed a large discrepancy between fatigue lives due to the characteristics of the material used. On the assumption that only one number of cycles corresponds to each pair of stress and strain amplitudes, the calculated fatigue lives should be equal or at least similar. Differences between calculated fatigue lives are probably an effect of different fatigue life characteristics used in the stress, strain and energy parameter based fatigue life assessment, since the distribution of the amplitudes was the same. Deliberations of the so-called compatibilities of fatigue characteristics can give us clearness to observed differences.

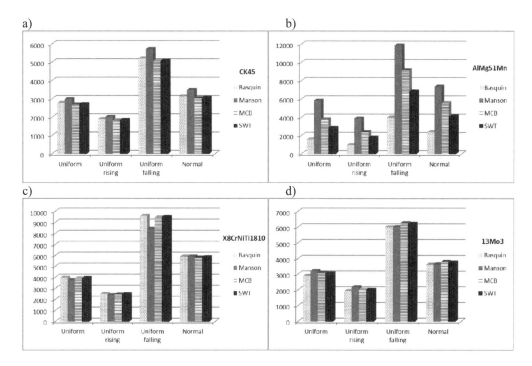

Fig. 5. Comparison of the calculated fatigue lives for four fatigue characteristics, four distributions of total strain amplitudes and four materials: (a) Ck45, (b) AlMg51Mn, (c) X8CrNiTi1810 and (d) 13Mo3

Let us consider a lack of so-called compatibility. It is the problem of equality of elastic and plastic parts in MCB and RO models. The equations of compatibility [17, 18]

$$n' = \frac{b}{c} = n'_{comp.}, \qquad (10)$$

$$K' = \frac{\sigma'_f}{(\varepsilon'_f)^{\frac{b}{c}}} = K'_{comp.}, \qquad (11)$$

are obtained by comparison of the expressions describing the elastic and plastic parts of the strain amplitude from the MCB and RO models. When the equality is not preserved, the fatigue life calculations give different values. It should be remembered especially in the case of fatigue life calculations for aluminium alloys where the differences are often big. Theoretical considerations concerning that problem can be found in [18]. Table 2 contain constants according to the selected material and those obtained for particular materials according to the equations of compatibility (10) and (11).

Suitable material tests and a good choice of the approximating equations should provide a correct description of fatigue properties of constructional materials. However, differences occurring in theoretical assumptions of the approximating formulas can influence the results of fatigue calculations.

Table 2. Material constants of four selected materials

Material	$\dfrac{b}{c} = n'_{comp.}$	n'	$\dfrac{\sigma'_f}{(\varepsilon'_f)^{\frac{b}{c}}} = K'_{comp.}$	K'
CK45	0.116	0.115	987.7	980
AlMg5,1Mn	0.156	0.0843	905	636.4
X8CrNiTi1810	0.416	0.421	5111	5234
15Mo3	0.178	0.18	878	893

References

[1] Mitchell, M.R.: Fundamentals of Modern Fatigue Analysis for Design, in: ASM Handbook, Ed. Steven R. Lampman, ASM International, Materials Park, 1996, pp. 229-249

[2] Rice, R.C.; Leis, B.N.; Berns, H.D.; Nelson, D.V.; Lingenfleser, D. and Mitchell, M.R.: Fatigue Design Handbook, SAE, Warrendale, 1988, 369 ps.

[3] Sonsino C.M.: Fatigue testing under variable amplitude loading, International Journal of Fatigue, 29(6), 2007, pp. 1080-1089

[4] Niesłony A.: Rain flow counting method, set of functions with user guide for use with MATLAB, Information on ⟨http://www.mathworks.com/matlabcentral/fileexchange/3026⟩

[5] Basquin O.H.: The exponential law of endurance tests, Am. Soc. Test. Mater. Proc., Vol. 10, 1910, pp. 625-630

[6] Manson S.S.: Fatigue: a complex subject – some simple approximation. Experimental Mechanics, Vol. 5, 1965, pp. 193-226

[7] Coffin L.F.: A study of the effect of cyclic thermal stresses on a ductile metal, Trans ASME, Vol. 76, 1954, pp. 931-950

[8] Manson S.S., Halford G.R.: Fatigue and Durability of Structural Materials, ASM International, Materials Park, Ohio, 2006, 456 ps.

[9] Kocańda S.: Fatigue Failure of Metals, Springer-Verlag GmbH, 1978, 388 ps.

[10] Macha E. and Sonsino C.M.: Energy criteria of multiaxial fatigue failure, Fatigue Fract Engng Mater Struct, 22, 1999, pp. 1053-1070

[11] Jahed H. and Varvani-Farahani A.: Upper and lower fatigue life limits model using energy-based fatigue properties, International Journal of Fatigue, Volume 28, Issues 5-6, 2006, pp. 467-473

[12] Smith K.N., Watson P. and Topper T.H.: A stress strain function for the fatigue of metals, J. Materials, Vol. 5, 1970, pp. 767-776

[13] ASTM Standard E606-92: Standard practice for strain-controlled fatigue testing. In: Annual book of ASTM standards, Vol. 03.01. ASTM; 1997, p. 523-37

[14] Bäumel A. and Seeger T.: Material Data for Cyclic Loading, Materials Science Monographs, 61, Elsevier Science Publishers, Amsterdam, 1990

[15] Ramberg W. and Osgood W.R.: Description of stress-strain curves by three parameters, Technical Note No. 902, National Advisory Committee for Aeronautics, Washington DC, 1943

[16] Plumtree A. and Abdel-Raouf H.A.: Cyclic stress–strain response and substructure, International Journal of Fatigue, Vol. 23, 2001, pp. 799-805

[17] Niesłony A., El Dsoki Ch., Kaufmann H., Krug P.: New method for evaluation of the Manson-Coffin-Basquin and Ramberg-Osgood equations with respect to compatibility, International Journal of Fatigue, Vol. 30, 2008, pp. 1967-1977

[18] Niesłony A., Kurek A., El Dsoki Ch. and Kaufmann H.: A study of compatibility between two classical fatigue curve models based on some selected structural materials, International Journal of Fatigue, 2011, Available online doi:10.1016/j.ijfatigue.2011.03.002

Keyword Index

A
Active Stabilization 187
Active Suspension System 161
Active Systems 187
Application of Design Methods 67
Applications of Robust Optimization 13
Automotive Industry 95

B
Behaviour Prediction Framework 3

C
Classification 33
Collaboration 55
Comfort Systems 75
Control of Uncertainty 33
Crashworthiness 145
Cumulative Damage 197

D
Data Uncertainties 13
Deep-Drawing 95
Design Criteria 23

E
Engineering Design 13
Evaluation of Robustness 75

F
Fatigue Characteristics 125
Fatigue Life 125, 145
Fatigue Life Assessment 197
Finite Element Analysis (FEA) 23
Flexibility 83
Fuzzy Number 23
Fuzzy Randomness 45

I
Imprecise Probability 45
Imprecision 45
Incremental Forming 115
Information Model 55
Interval Number 23

J
Joining 115

L
Load-Monitoring 187

M
Managing Uncertainty 3
Mechatronic Systems 75
Metal Forming 83
Metal Spinning 115
Methods 33
Model of Uncertainty 33
Model Predictive Control 3
Monte Carlo Simulation 177

N
Nonlinear Programming 13

O
Out-Of-Parallelism 125

P
Performance Verification 177
Probability Density Function 23
Process Chains 133
Process Model 133
Process Simulation 103

R
Railway Design 177
Reaming 103
Reliability Analysis 45
Robust Design 67, 75, 161
Robust Optimization 13
Robustness 13
Rotary Swaging 115

S
Servo Press 83
Shape Optimization 13
Sheet Metal 95
Smart Structures 115
Strain-Life 197
Stress-Life 197
Structural Complexity Management 3

T
Tolerance Management 67, 95

U

Uncertainty 33, 45, 55, 83, 103, 133, 145, 187
Uncertainty in Railway Vehicle 177

V

Variable Amplitude Loading 197
Variation Prediction 95
Vehicle Subsystems 145
Vibration Control 161
Virtual Testing 145

Author Index

A
Abele, E. 103
Anderl, R. 55

B
Bedarff, T. 161
Bohn, A. 33
Bohn, M. 67, 75, 95
Brenneis, M. 33, 115
Bruni, S. 177

C
Calmano, S. 83

D
d'Ippolito, R. 145
Desmet, W. 145
Donders, S. 145

E
Ederer, T. 83
Eifler, T. 133
Engelhardt, R. 33
Enss, G.C. 133, 187

F
Farkas, L. 145

G
Graf, W. 45
Groche, P. 83, 115

H
Hack, M. 145
Hanselka, H. 133, 187
Hauer, T. 103
Haydn, M. 55, 103, 133

K
Kloberdanz, H. 33
Koenen, J.F. 33, 187
Kraft, M. 83
Kurek, A. 197
Kurek, M. 125

L
Łagoda, T. 125
Lorenz, U. 83

M
Marjanović, D. 3
Mathias, J. 161
Mazzola, L. 177
Moens, D. 23, 145
Mosch, L. 55, 133

N
Niesłony, A. 197

O
Ondoua, S. 55, 187
Osman, K. 3

P
Pelz, P.F. 161
Platz, R. 133, 187

S
Schmitt, S. 83
Sichau, A. 13
Sickert, J.U. 45
Sprenger, A. 55
Stanković, T. 3
Steinle, P. 67, 95
Štorga, M. 3

T
Türk, M. 115

U
Ulbrich, S. 13

V
Van der Auweraer, H. 145
Vandepitte, D. 23

W
Wuttke, F. 75, 95